Modern Projective Geometry

Modern Projective Geometry

Robert J. Bumcrot
Hofstra University

HOLT, RINEHART and WINSTON, INC.
New York Chicago San Francisco Atlanta Dallas
Montreal Toronto London Sydney

To my wife Dolores

Library of Congress Catalog Card Number: 69-10905

2721652

Printed in the United States of America

1 2 3 4 5 6 7 8 9

Preface

This is a textbook for a serious advanced undergraduate or first-year graduate course in projective geometry. We assume that the reader has had two or three years of college mathematics, including a general introductory course in modern algebra and a course in finite-dimensional linear algebra over an arbitrary field. Although Chapter I contains everything needed to make the book self-contained, a reader who is not already familiar with the first five or six sections of this chapter may find many parts of the later chapters difficult.

Several different courses can be based on topics covered in this book. A basic course should concentrate on Chapters II, III and IV, which require only the first half of Chapter I. These chapters contain the material on projective spaces most often needed in other parts of mathematics, as well as the necessary background for the two branches of current geometrical research discussed in the final chapters. It should be noted that Chapter IV can be presented before Chapter III.

Chapter V is a brief introduction to the analytic geometry of curves, surfaces, and their higher-dimensional analogs, culminating in the first few faltering steps toward that branch of algebra called algebraic geometry. Chapter VI contains a good many of the most important basic results in the theory of finite projective planes. Given sufficient group-theoretic background, the reader can proceed from this chapter to the current literature.

More than two hundred exercises have been provided. They range from straightforward computations and verifications to complicated extensions of the theory and results on concepts not studied in the text. We have tried to indicate it whenever the solution of an exercise may involve knowledge not possessed by all our readers. A few of the more difficult exercises are starred.

A word should be said about our occasional use of the pseudo-mathematical terms *synthetic* and *analytic*. By synthetic geometry we mean geometry in which the definitions, theorems, and proofs are given without recourse to algebra or analysis. By analytic geometry we understand any sort of geometry that is not purely synthetic.

For many stimulating courses, discussions, and communications, not all limited to the field of projective geometry, I wish to thank A. A. Albert, A. Barlotti, L. M. Blumenthal, S.-C. Chern, J. Hirschfeld, D. R. Hughes, I. Kaplansky, V. Klee, F. A. Sherk, A. Weir, J. L. Zemmer, and the members of the Ohio State and University of Sussex geometry seminars.

Special thanks are due to A. Cronheim, H. Lenz, L. F. Meyers, and J. C. D. S. Yaqub, each of whom read all or part of the manuscript.

Finally I wish to express my appreciation to the secretaries of the division of mathematics of the University of Sussex, for their typing of the manuscript, and to my wife, for her help throughout its preparation.

R. J. B.

Hempstead, New York
October 1968

Contents

Modern Projective Geometry

I

Preliminaries

This chapter contains most of the results from other branches of mathematics that will be required. The theorems and definitions should be largely familiar to the reader described in the preface; unfamiliar results should be easily understandable, although not necessarily easily provable. We have provided proofs for the few nonstandard results needed. Proofs of the other theorems may be supplied by the reader, if he wishes to work them out, or they may be found in various textbooks.

Sections 1 through 5 present fundamental information. Later sections will be cited in the introductions to the chapters to which they are pertinent.

1. BASIC NOTATION

A reference to Item n refers to Item n of the same section; $s.n$ refers to Item n of Section s of the same chapter; $R.s.n$ (where R is a Roman numeral) refers to Item n of Section s of Chapter R.

A reference [n] refers to Item n of the bibliography.

The end of a proof or partial proof is denoted by ▎.

If p and q are statements, then '$p \rightarrow q$' means 'p implies q' or 'if p, then q.' '$p \leftrightarrow q$' or 'p iff q' means 'p is equivalent to q' or 'p if and only if q.' The statement 'p or q' includes the possibility 'p and q.'

2. SET THEORY

We shall adopt a naïve approach to the theory of sets. A *set* S is any collection of objects, called the *elements* of S. If x is an element of S, we write $x \in S$; if x is not an element of S, we write $x \notin S$.

A set A is a *subset* of a set B, written either $A \subset B$ or $B \supset A$, if for all x, $x \in A \rightarrow x \in B$. A and B are *equal*, written $A = B$, if $A \subset B$ and $B \subset A$. If A and B are not equal, we write $A \neq B$. A is a *proper subset* of B, written $A \subsetneq B$, if $A \subset B$ and $A \neq B$. If A is not a subset of B, we write $A \not\subset B$. We say A and B are *noncomparable* if $A \not\subset B$ and $B \not\subset A$.

If the set A consists of all those elements of a set S that have a certain property p, we write $A = \{x \in S \mid p\}$ or, if S is understood, $A = \{x \mid p\}$.

The *union*, *intersection*, and *difference* of two sets A, B are defined and written as follows: $A \cup B = \{x \mid x \in A \text{ or } x \in B\}$, $A \cap B = \{x \mid x \in A \text{ and } x \in B\}$, $A - B = \{x \mid x \in A \text{ and } x \notin B\}$.

The set with no elements—the *empty set*—is denoted by $\emptyset$. Sets A and B are *disjoint* if $A \cap B = \emptyset$.

Suppose we have a set S_i for each i in a nonempty *indexing set* I. The *union* and *intersection* of the family $\{S_i \mid i \in I\}$ are defined and written as follows: $\bigcup_{i \in I} S_i = \{x \mid x \in S_i \text{ for some } i \in I\}$, $\bigcap_{i \in I} S_i = \{x \mid x \in S_i \text{ for all } i \in I\}$.

Elements $x_1, x_2, \cdots$ are *distinct* if $x_i = x_j \rightarrow i = j$. A *finite* set is one that has only a finite number of distinct elements. If the finite set A has exactly n elements, where n is a nonnegative integer, we write $|A| = n$. If $x_1, x_2, \cdots, x_n$ are the elements of A, we write $A = \{x_1, x_2, \cdots, x_n\}$. If A and B are finite sets, then $|A \cup B| + |A \cap B| = |A| + |B|$. A and B are disjoint iff $|A \cup B| = |A| + |B|$.

The *ordered pair* (a, b) of elements a, b is the set $\{\{a\}, \{a, b\}\}$. We have $(a, b) = (c, d)$ iff $a = c$ and $b = d$. The *Cartesian product* $A \times B$ of two sets A, B is $\{(a, b) \mid a \in A \text{ and } b \in B\}$. We define the *ordered triple* (a, b, c) to be $((a, b), c)$ and the Cartesian product $A \times B \times C$ to be $(A \times B) \times C$. In general, for any n we define the *ordered n-tuple* $(x_1, x_2, \cdots, x_n)$ to be $((x_1, \cdots, x_{n-1}), x_n)$, and similarly for the Cartesian product $A_1 \times A_2 \times \cdots \times A_n$. If $A_1 = A_2 = \cdots = A_n = A$, we write A^n for $A_1 \times A_2 \times \cdots \times A_n$.

A *relation* from a set A to a set B is a subset R of $A \times B$. We write $x\,R\,y$ for $(x, y) \in R$. The *domain* and *range* of R are $\operatorname{dom} R = \{x \in A \mid x\,R\,y \text{ for some } y \in B\}$, $\operatorname{ran} R = \{y \in B \mid x\,R\,y \text{ for some } x \in A\}$. A *function* or *mapping* from A to B is a relation f from A to B such that $x\,f\,y$ and $x\,f\,y' \rightarrow y = y'$. We write $f(x) = y$, or sometimes $y = xf$, for $x\,f\,y$. (Be prepared for either notation!) The symbols '$f\colon A \rightarrow B$' mean 'f, which is a function from A to B'. $f\colon A \rightarrow B$ does not imply that $\operatorname{dom} f = A$ or that $\operatorname{ran} f = B$.* If $x \in \operatorname{dom} f$, then $f(x)$ is called the *image* of x under f.

A function $f\colon A \rightarrow B$ is *defined on A* if $\operatorname{dom} f = A$; f is *onto B* if $\operatorname{ran} f = B$. f is *one–one* if, for all x and x' in $\operatorname{dom} f$, $f(x) = f(x') \rightarrow x = x'$. A one–one function defined on A onto B is called a *one–one correspondence* between A and B.

If R is a relation from A to B and $C \subset A$ and $D \subset B$, then the *restriction* of R to C and D is $R \cap (C \times D)$. If $S \subset A$, we define $R(S)$ to be $\{y \in B \mid x\,R\,y \text{ for some } x \text{ in } S\}$. In case R is a function f, this becomes $f(S) = \{f(x) \mid x \in S\}$.

The *inverse* of any relation R is the relation $R^{-1} = \{(y, x) \mid (x, y) \in R\}$. The inverse of a function f is a relation; it is a function iff f is one–one.

A relation R from a set A to itself is an *equivalence relation* on A if for all x, y, z in A the following conditions hold: $x\,R\,x$; $x\,R\,y \rightarrow y\,R\,x$; $x\,R\,y$ and $y\,R\,z \rightarrow x\,R\,z$. For each x in A we define the *equivalence class* of an equivalence relation R on A to be $R(x) = \{y \in A \mid x\,R\,y\}$. Then

---

* Confusion with the $\rightarrow$ of logical implication is most unlikely.

$$y \in R(x) \leftrightarrow R(x) = R(y). \tag{*}$$

A family of nonempty subsets of a set A is a *partition* of A if each element of A is in one and only one member of the family. By (*) the family of all equivalence classes of an equivalence relation on A is a partition of A. Conversely, let P be a partition of A. Define the relation R as follows: $x \, R \, y$ iff x and y are in the same member of P. Then R is an equivalence relation on A, and the equivalence classes of R are just the members of P. This establishes a one–one correspondence between partitions of A and equivalence relations on A.

A *partially ordered set* is a set A together with a relation $\leq$ from A to A that satisfies the following conditions for all x, y, z in A: $x \leq x$; $x \leq y$ and $y \leq x \rightarrow x = y$; $x \leq y$ and $y \leq z \rightarrow x \leq z$. *Example:* The family of all subsets of a given set S together with the relation $\subset$ is a partially ordered set.

A *chain* in a partially ordered set $(A, \leq)$ is a subset C of A such that, for all x and y in C, $x \leq y$ or $y \leq x$.

An element a of a partially ordered set $(A, \leq)$ is an *upper* (*lower*) *bound* of a subset B of A if, for all x in B, $x \leq a$ (respectively, $a \leq x$). a is the (necessarily unique) *least upper* (*greatest lower*) *bound* of B if it is an upper (lower) bound of B and also a lower (upper) bound of the set of all upper (lower) bounds of B. An upper (lower) bound of A itself is called a *maximal* (*minimal*) element. In the example above, S is a maximal element, $\emptyset$ is a minimal element, and if D and E are subsets of S, then their least upper (greatest lower) bound is $D \cup E$ $(D \cap E)$.

We need the following form of one of the axioms of set theory, known as the *Minimum Principle*. Let $(A, \leq)$ be a nonempty partially ordered set. If every chain in A has a lower bound in A, then A contains at least one minimal element.

3. SOME ALGEBRAIC SYSTEMS

A mathematical system usually consists of an underlying set together with relations. When no confusion is likely to result, we shall refer to a given system by its underlying set only.

A *binary operation* on a set A is a function $b\colon A \times A \rightarrow A$ defined on $A \times A$. $b((x, y))$ is usually written xy, $x + y$, $x \circ y$, or similarly. A *binary system* is a set A together with a binary operation on A. *Ternary* operation and system are defined analogously.

An *identity element* of a binary system $(A, \circ)$ is an element e of A such that $e \circ x = x \circ e = x$ for all x in A. e is unique, if it exists at all, and is denoted by 1 if the operation is written multiplicatively and by 0 if it is written additively.

A *loop* is a binary system $(L, \cdot)$ with identity 1 such that for every a and b in L there is a unique u in L such that $au = b$ and there is a unique v in L such that $va = b$. (The symbol $\cdot$ is omitted in writing.)

A *group* is a loop $(G, \cdot)$ in which, for all a, b, c, $a(bc) = (ab)c$. An *Abelian* group is a group in which $ab = ba$ for all a, b.

A *ring* is a set R together with two binary operations $+$ and $\cdot$ such that $(R, +)$ is an Abelian group and, for all a, b, c: $a(bc) = (ab)c$, $a(b + c) = ab + ac$, and $(b + c)a = ba + ca$. A ring is *commutative* if $ab = ba$ for all a, b. A ring *has identity* if $(R, \cdot)$ has an identity element.

Let $(R, +, \cdot)$ be a ring with identity 1. If n is a positive integer, $n \cdot 1$ means $1 + 1 + \cdots + 1$ (n 1's). If for all n we have $n \cdot 1 \neq 0$ (the additive identity of R), we say R has *characteristic zero*. Otherwise, the smallest n such that $n \cdot 1 = 0$ is called the *characteristic* of R.

An *integral domain* (with identity) is a commutative ring with identity in which, for all a and b, $ab = 0 \rightarrow a = 0$ or $b = 0$.

A *division ring* is a ring $(Q, +, \cdot)$ such that $(Q - \{0\}, \cdot)$ is a group. A *field* is a commutative division ring. The fields of real and complex numbers are denoted R and C.

4. VECTOR SPACES

A *vector space* over a field F is an Abelian group $(V, +)$ together with a function from $F \times V$ to V defined on $F \times V$, in which the following conditions hold for all a, b in F and $\mathbf{u}, \mathbf{v}$ in V. (We write $a\mathbf{u}$ for the image in V of $(a, \mathbf{u})$ under the function.) $(a + b)\mathbf{u} = a\mathbf{u} + b\mathbf{u}$, $a(b\mathbf{u}) = (ab)\mathbf{u}$, $1\mathbf{u} = \mathbf{u}$, $a(\mathbf{u} + \mathbf{v}) = a\mathbf{u} + a\mathbf{v}$. Confusion over the double use of additive and multiplicative notation is unlikely. Elements of V are *vectors;* elements of F are *scalars.*

A set U of vectors is *independent* if for all finite subsets $\{\mathbf{u}_1, \mathbf{u}_2, \cdots, \mathbf{u}_n\}$ of U and scalars $a_1, a_2, \cdots, a_n$: $a_1\mathbf{u}_1 + a_2\mathbf{u}_2 + \cdots + a_n\mathbf{u}_n = \mathbf{0} \rightarrow a_1 = a_2 = \cdots = a_n = 0$. U is *dependent* if it is not independent.

A nonempty subset U of a vector space V is a *subspace* of V if, for all $\mathbf{u}, \mathbf{u}'$ in U and scalars a, $a\mathbf{u} + \mathbf{u}'$ is in U. A subspace, together with the restricted operations of its containing space, is itself a vector space.

A subset U of a vector space V *spans* V if given any $\mathbf{v}$ in V there are vectors $\mathbf{u}_1, \mathbf{u}_2, \cdots, \mathbf{u}_n$ in U and scalars $a_1, a_2, \cdots, a_n$ such that $\mathbf{v} = a_1\mathbf{u}_1 + a_2\mathbf{u}_2 + \cdots + a_n\mathbf{u}_n$. A subset B of V is a *basis* for V if B is independent and spans V. It follows from the Minimum Principle that every vector space has a basis.

If B is a basis for a vector space V, then for every nonzero $\mathbf{v}$ in V there are *unique* vectors $\mathbf{b}_1, \mathbf{b}_2, \cdots, \mathbf{b}_n$ in B and nonzero scalars $v_1, v_2, \cdots, v_n$ such that $\mathbf{v} = v_1\mathbf{b}_1 + v_2\mathbf{b}_2 + \cdots + v_n\mathbf{b}_n$.

If some basis of a vector space V consists of a finite number n of vectors, then every basis of V will contain exactly n vectors. We then say V is n-dimensional and write $\dim V = n$.

Every independent set of vectors in a vector space V is contained in a basis of V. Every set of vectors that spans V contains a basis of V.

Let $\{\mathbf{b}_1, \mathbf{b}_2, \cdots, \mathbf{b}_n\}$ be a basis for an n-dimensional vector space V over a field F. Every $\mathbf{v}$ in V can be written uniquely in the form $v_1\mathbf{b}_1 + v_2\mathbf{b}_2 + \cdots + v_n\mathbf{b}_n$, where some v_i's may be 0. The function $\mathbf{v} \to (v_1, v_2, \cdots, v_n)$ so defined is a one–one correspondence between V and F^n. If $a \in F$ then $a\mathbf{v} \to (av_1, av_2, \cdots, av_n)$. Also if $u \to (u_1, u_2, \cdots, u_n)$ then $\mathbf{u} + \mathbf{v} \to (u_1 + v_1, u_2 + v_2, \cdots, u_n + v_n)$. With these operations, F^n is a vector space over F. We say V and F^n are *isomorphic* under this correspondence. Thus every n-dimensional vector space over F is isomorphic to F^n.

If U, U' are finite-dimensional subspaces of a vector space V, then so are $U \cap U'$ and $U + U'$, which is defined as $\{\mathbf{u} + \mathbf{u}' \mid \mathbf{u} \in U \text{ and } \mathbf{u}' \in U'\}$. We have $\dim (U + U') + \dim (U \cap U') = \dim U + \dim U'$.

Let B be a basis of a vector space V over a field F. If $\mathbf{u}$ and $\mathbf{v}$ are in V, then there are vectors $\mathbf{b}_1, \mathbf{b}_2, \cdots, \mathbf{b}_n$ in B and scalars $u_1, u_2, \cdots, u_n, v_1, v_2, \cdots, v_n$ in F (some perhaps zero) such that $\mathbf{u} = u_1\mathbf{b}_1 + u_2\mathbf{b}_2 + \cdots + u_n\mathbf{b}_n$ and $\mathbf{v} = v_1\mathbf{b}_1 + v_2\mathbf{b}_2 + \cdots + v_n\mathbf{b}_n$. The *dot product* of $\mathbf{u}$ and $\mathbf{v}$ is the scalar $\mathbf{u} \cdot \mathbf{v} = u_1v_1 + u_2v_2 + \cdots + u_nv_n$. This scalar depends on B. If $V = F^n$ and the *usual basis* $(1, 0, \cdots, 0), (0, 1, \cdots, 0), \cdots, (0, \cdots, 0, 1)$ is used, then $(u_1, u_2, \cdots, u_n) \cdot (v_1, v_2, \cdots, v_n) = u_1v_1 + u_2v_2 + \cdots + u_nv_n$. Whenever dot products are used in F^n, we always assume they are taken with respect to the usual basis, unless some other assumption is specifically stated.

Let V be a vector space, with some fixed basis with respect to which we take dot products. Two vectors in V are *orthogonal* if their dot product is zero. Two subspaces U, U' of V are orthogonal if every vector in U is orthogonal to every vector in U'. Every subspace U of V has a unique *orthogonal complement* subspace $U^\perp$ such that U is orthogonal to $U^\perp$ and $U + U^\perp = V$. If V is finite-dimensional we have $\dim U + \dim U^\perp = \dim V$.

5. MATRICES AND LINEAR TRANSFORMATIONS

An $n \times m$ *matrix* A over a ring R is an array of n rows and m columns of elements of R,

$$A = \begin{bmatrix} a_{11} & a_{12} & \cdots & a_{1m} \\ a_{21} & a_{22} & \cdots & a_{2m} \\ \vdots & & & \\ a_{n1} & a_{n2} & \cdots & a_{nm} \end{bmatrix}.$$

If $A = (a_{ij})$, $B = (b_{ij})$ are $n \times m$ matrices, their *sum* $A + B$ is the $n \times m$ matrix (c_{ij}), where $c_{ij} = a_{ij} + b_{ij}$. If $A = (a_{ij})$ is $n \times m$ and $B = (b_{jk})$ is $m \times p$, their *product* AB is the $n \times p$ matrix (c_{ik}), where $c_{ik} = a_{i1}b_{1k} + a_{i2}b_{2k} + \cdots + a_{im}b_{mk}$. The set of all $n \times n$ matrices over R forms a ring $M_n(R)$ under these operations. If R has identity 1 then $M_n(R)$ has identity $I = (\delta_{ij})$, where δ_{ij} is 1 if $i = j$ and is 0 if $i \neq j$.

The *transpose* of the $n \times m$ matrix $A = (a_{ij})$ is the $m \times n$ matrix A^T, defined as (a'_{ij}), where $a'_{ij} = a_{ji}$. We have $(A + B)^T = A^T + B^T$ and $(AB)^T = B^T A^T$ whenever either side of the equality is defined. The one-row matrix $(a_1 a_2 \cdots a_n)$ and the one-column matrix $(a_1 a_2 \cdots a_n)^T$ over a field F are often identified with the vector $\mathbf{a} = (a_1, a_2, \cdots, a_n)$ in F^n. If A is an $n \times n$ matrix over F, $\mathbf{a}A$ means $(a_1 a_2 \cdots a_n)A$, whereas $A\mathbf{a}$ means $A(a_1 a_2 \cdots a_n)^T$.

An $n \times n$ matrix A is *nonsingular* if there exists an $n \times n$ *inverse matrix* A^{-1} such that $AA^{-1} = A^{-1}A = I$. The set of all nonsingular $n \times n$ matrices over a field forms a group under multiplication. If a is an element of a field, the $n \times n$ *scalar matrix* aI is the matrix $(a\delta_{ij})$. The set of all nonzero $n \times n$ scalar matrices over a field forms an Abelian group under matrix multiplication. If A is a matrix and a a scalar, aA means $(aI)A$.

The *determinant* det A of the 1×1 matrix $A = (a)$ over an integral domain is defined to be just a. Proceeding inductively, the determinant of the $n \times n$ matrix $A = (a_{ij})$ is defined to be $\det A = a_{11} \det A_{11} - a_{12} \det A_{12} + \cdots + (-1)^{n+1} a_{1n} \det A_{1n}$, where A_{ij} is A with its ith row and jth column erased. The basic properties of determinants are as follows: $\det A = \det A^T$. If two rows of A are switched, det A changes sign. If one row of A is multiplied through by a number k, det A is multiplied by k. If a multiple of one row of A is added to another row, det A does not change. $\det(AB) = (\det A)(\det B)$.

An $n \times n$ matrix A over a field is nonsingular if and only if $\det A \neq 0$.

Let V and W be vector spaces over a field F. A *linear transformation* from V to W is a function $f: V \rightarrow W$ such that for all $\mathbf{v}$, $\mathbf{v}'$ in V and a in F, $f(a\mathbf{v} + \mathbf{v}') = af(\mathbf{v}) + f(\mathbf{v}')$. If $V = W$, then we call f a linear transformation of V.

Let A be an $n \times n$ matrix over a field F. The functions $f(\mathbf{x}) = \mathbf{x}A$ and $g(\mathbf{x}) = A\mathbf{x}$ are each linear transformations of F^n. The following four subspaces of F^n have the same dimension, called the *rank* of A: $f(F^n)$, $g(F^n)$, the subspace spanned by the rows of A regarded as vectors, the subspace spanned by the columns of A regarded as vectors.

If $f: V \rightarrow W$ is a linear transformation, the sets $R_f = f(V)$ and $N_f = f^{-1}(\mathbf{0})$ are subspaces of W and V, respectively, called the *range space* and *null space* of f. If V is finite-dimensional, so are R_f and N_f, and then $\dim V = \dim R_f + \dim N_f$. f is one–one iff $\dim N_f = 0$, that is, iff $N_f = \{\mathbf{0}\}$.

Let f be a linear transformation of a finite-dimensional vector space V and let $B = \{\mathbf{b}_1, \mathbf{b}_2, \cdots, \mathbf{b}_n\}$ be a basis of V. For each $i = 1, 2, \cdots, n$ there are unique scalars $a_{i1}, a_{i2}, \cdots, a_{in}$ such that $f(\mathbf{b}_i) = \sum_{j=1}^{n} a_{ij}\mathbf{b}_j$. The $n \times n$ matrix $A_f = (a_{ij})$ is called the *matrix* of f *relative to B*. If V is F^n and B is the usual basis, then $f(\mathbf{x}) = A_f\mathbf{x}$ for all $\mathbf{x}$. The following four statements are equivalent: f is one–one, f is onto, the matrix of f relative to some basis is nonsingular, the matrices of f relative to all bases are nonsingular. If A and A' are matrices of f relative to different bases, then A is *similar* to A'. This means that there exists a nonsingular matrix C such that $A' = CAC^{-1}$. Similarity of $n \times n$ matrices is an equivalence relation.

Consider a system of n linear homogeneous equations in n unknowns over a field F:

$$\sum_{j=1}^{n} a_{ij}x_j = 0, \qquad i = 1, 2, \cdots, n. \tag{*}$$

Let $\mathbf{a}_i = (a_{i1}, a_{i2}, \cdots, a_{in})$, $\mathbf{x} = (x_1, x_2, \cdots, x_n)$, $A = (a_{ij})$, $f(\mathbf{x}) = A\mathbf{x}$. Many results of linear algebra depend on the fact that the following three sets are equal: the set of all solutions of (*) (regarded as vectors); the orthogonal complement in F^n of the subspace spanned by $\mathbf{a}_1, \mathbf{a}_2, \cdots, \mathbf{a}_n$; the null space of f.

6. FIELDS

See Section 3 for definitions.

The characteristic of any field is either zero or a prime number. Every finite field has prime characteristic. If F is a finite field of characteristic p, then $|F| = p^n$ for some positive integer n. For each prime p and positive integer n there is a unique finite field F, of characteristic p, with $|F| = p^n$. This field is usually denoted by $GF(p^n)$ (Galois field of order p^n).

An *automorphism* of a field F is a one–one correspondence α between F and itself such that, for all x and y in F, $\alpha(x + y) = \alpha(x) + \alpha(y)$ and $\alpha(xy) = \alpha(x)\alpha(y)$. The only automorphism of the field R of real numbers is the *identity automorphism I* such that $I(x) = x$ for all x in R.

An *ordered field* is a field F together with a partial ordering $\leq$ such that the following conditions are satisfied for all x, y, z in F: either $x \leq y$ or $y \leq x$; if $x \leq y$, then $x + z \leq y + z$; if $x \leq y$ and $0 \leq z$, then $xz \leq yz$. The fields of real and rational numbers, with the usual interpretation of $\leq$, are examples of ordered fields. There is no partial ordering of the field of complex numbers that can make it an ordered field.

It was shown by J. H. MacLagan-Wedderburn that every finite division ring is a field. This result was later strengthened by E. Artin and M. Zorn, who showed that every finite *alternative division ring* is a field. An alternative division ring is a set D together with two binary operations $+$ and $\cdot$ satisfying: $(D, +)$ is an Abelian group; for all x, y, z, $(x + y)z = xz + yz$ and $z(x + y) = zx + zy$; there is a multiplicative identity 1; each $x \neq 0$ has a (necessarily unique) inverse x^{-1} such that $xx^{-1} = x^{-1}x = 1$; for all nonzero x and y, $x^{-1}(xy) = y$ and $(xy)y^{-1} = x$.

7. POLYNOMIALS

We assume that the reader knows what a *polynomial*, in one or several variables over a field F, is and that he knows how to add and multiply polynomials. These manipulations obviously still make sense if the field of coefficients F is replaced by a commutative ring R.

The set $R[x_1, x_2, \cdots, x_n]$ of all polynomials in n variables over the ring R is itself a ring under polynomial addition and multiplication. The abbreviation $\mathbf{x}$ for $x_1, x_2, \cdots, x_n$ is employed when no confusion is likely. If R is an integral domain, so is $R[\mathbf{x}]$. A polynomial in n variables over an integral domain D can be written as a polynomial in any one of the variables x_i over the integral domain $D[x_1, \cdots, x_{i-1}, x_{i+1}, \cdots, x_n]$. For example, $3x_1^4x_2 + x_1x_2^3 + 4x_1^2x_2^3 + 5x_1^5 = (4x_1^2 + x_1)x_2^3 + (3x_1^4)x_2 + (5x_1^5) \in D[x_2]$, where $D = Z[x_1]$ and Z is the integral domain of integers. By this means, calculations with polynomials in several variables can be reduced to a series of calculations with polynomials in one variable.

The *formal derivative* of the polynomial $p(x) = a_nx^n + \cdots + a_1x + a_0$ is the polynomial $p'(x) = na_nx^{n-1} + \cdots + a_1$. The formal derivative of a constant polynomial is zero. The *formal partial derivative* of a polynomial $p(\mathbf{x}) = p(x_1, \cdots, x_n)$ with respect to the ith variable x_i is just the derivative of the polynomial written as a polynomial in x_i only. Thus, the partial derivative with respect to x_2 of the example polynomial in the preceding paragraph is $3(4x_1^2 + x_1)x_2^2 + (3x_1^4) = 12x_1^2x_2^2 + 3x_1^4$. The partial derivative of $p(\mathbf{x})$ with respect to x_i is denoted either $p_{x_i}(\mathbf{x})$ or $p_i(\mathbf{x})$. Repeating the above process, we may define the second, third, $\cdots$ derivatives of a polynomial in one or several variables. The rth derivative of $p(x)$ is denoted $p^{(r)}(x)$. The derivative of $p(x_1, x_2)$ twice with respect to x_1 and once with respect to x_2, for example, is denoted either $p_{x_1^2x_2}(\mathbf{x})$ or $p_{112}(\mathbf{x})$. The order in which these differentiations are performed is immaterial.

The following rules of differential calculus hold for formal derivatives (ordinary or partial): $(p(x) + q(x))' = p'(x) + q'(x)$; $(p(x)q(x))' = p(x)q'(x) + p'(x)q(x)$; $(p(q(x)))' = p'(q(x))q'(x)$. We require only the following case of the general chain rule. Let $p(x, y)$ be a polynomial, let a, b, c, d be constants, and let $f(t)$ be the polynomial $p(a + ct, b + dt)$. Then $f^{(r)}(0)$, the rth derivative of f evaluated at $t = 0$, is equal to

$$p_{x^r}(a, b)c^r + \binom{r}{1} p_{x^{r-1}y}(a, b)c^{r-1}d + \cdots$$
$$+ \binom{r}{i} p_{x^{r-i}y^i}(a, b)c^{r-i}d^i + \cdots + p_{y^r}(a, b)d^r,$$

where

$$\binom{r}{i} = \frac{r!}{i!(r-i)!}.$$

Let

$$ax_1^{e_1}x_2^{e_2} \cdots x_n^{e_n}$$

be a term of a polynomial $p(\mathbf{x})$. (Here a is a nonzero constant and the e_i are nonnegative integers.) The *degree* of this term is the integer $e_1 + e_2 + \cdots + e_n$; the degree of $p(\mathbf{x})$ is the largest of the degrees of its terms.

We say a polynomial $p(\mathbf{x})$ *divides*, or is a *factor* of, a polynomial $q(\mathbf{x})$ if $q(\mathbf{x}) = p(\mathbf{x})r(\mathbf{x})$ for some polynomial $r(\mathbf{x})$. $p(\mathbf{x})$ is *prime* if its only nonconstant factors are those of the form $kp(\mathbf{x})$, where k is a constant. Every polynomial is, in an essentially unique way, the product of its prime factors.

Let $p(x) = a_nx^n + \cdots + a_0$ and $q(x) = b_mx^m + \cdots + b_0$ be polynomials over an integral domain D. We assume $a_n \neq 0$, $b_m \neq 0$. (Remember that p, q may really be polynomials in several variables besides x.) The *resultant* $R(p, q)$ of p and q is the determinant of the following $n + m$ by $n + m$ matrix.

$$\left[\begin{array}{ccccccc}
a_0 & a_1 & \cdots & a_n & 0 & \cdots & 0 \\
0 & a_0 & \cdots & a_{n-1} & a_n & \cdots & 0 \\
\vdots & & & & & & \vdots \\
0 & \cdots & a_0 & \cdot & \cdot & \cdot & a_n \\
b_0 & b_1 & \cdots & b_m & 0 & \cdots & 0 \\
0 & b_0 & \cdots & b_{m-1} & b_m & \cdots & 0 \\
\vdots & & & & & & \vdots \\
0 & \cdots & & b_0 & & \cdots & b_m
\end{array}\right]
\begin{array}{l} \left.\begin{array}{l}\\ \\ \\ \\ \end{array}\right\} m \text{ rows} \\ \left.\begin{array}{l}\\ \\ \\ \\ \end{array}\right\} n \text{ rows}\end{array}$$

The resultant of two polynomials is zero if and only if the polynomials have a common nonconstant factor. The *discriminant* of $p(x)$ is the resultant of $p(x)$ and its formal derivative. If the integral domain D has characteristic zero, then the discriminant of $p(x)$ is zero iff $p(x)$ has a repeated nonconstant factor, that is, $p(x) = q^2(x)r(x)$ where q is of positive degree.

Let $p(x)$ be a polynomial over a field F of characteristic zero. An element a of F is an *r-fold root* of $p(x)$ if $p(x) = (x - a)^r q(x)$, where $q(x)$ is a polynomial over F and $q(a) \neq 0$. a is also called a root of *multiplicity r*. a is an r-fold root of p iff $p(a) = p'(a) = \cdots = p^{(r-1)}(a) = 0$ and $p^{(r)}(a) \neq 0$. If p is of degree n, it has at most n roots in F, where an r-fold root is counted r times.

A root of a polynomial $p(\mathbf{x})$ in n variables over a field F is an n-tuple $\mathbf{a}$ of elements of F such that $p(\mathbf{a}) = 0$.

A field K is *algebraically closed* if every nonconstant polynomial over K has a root in K. The field C of complex numbers is algebraically closed. Let $p(\mathbf{x})$, $q(\mathbf{x})$ be polynomials over an algebraically closed field K. If every root of p is a root of the same or greater multiplicity of q, then p divides q. If p and q have the same roots with the same multiplicities, then $p(\mathbf{x}) = kq(\mathbf{x})$ for some nonzero constant k.

8. HOMOGENEOUS POLYNOMIALS

A polynomial $p(\mathbf{x})$ in one or several variables is called *homogeneous* if each of its terms has the same degree. $p(\mathbf{x})$ is homogeneous of degree k iff

$$p(t\mathbf{x}) = t^k p(\mathbf{x}) \qquad (*)$$

for all t, where $t\mathbf{x}$ means $tx_1, tx_2, \cdots, tx_n$. Differentiating (*) with respect to t and then setting $t = 1$, we have *Euler's theorem:* If $p(\mathbf{x})$ is a homogeneous polynomial of degree k over a commutative ring with identity, then

$$\sum_{i=1}^{n} p_i(\mathbf{x}) = kp(\mathbf{x}).$$

All factors and all derivatives of a homogeneous polynomial are homogeneous. Every homogeneous polynomial $p(\mathbf{x})$ has an essentially unique factorization $p(\mathbf{x}) = p^1(\mathbf{x})p^2(\mathbf{x}) \cdots p^n(\mathbf{x})$, where the $p^i(\mathbf{x})$ are prime homogeneous polynomials. (The numbers are superscripts, not exponents.)

Let $R(x_0, x_1)$ be a homogeneous polynomial of degree n over a field F. The roots of $R(x_0, x_1) = 0$ are often grouped into equivalence classes called *ratios*. A ratio $a{:}b$ (a, b not both zero) is $\{(ta, tb) \mid t \in F\}$. If any (t_0a, t_0b), $t_0 \neq 0$, is a root of R, then all (ta, tb) are roots of R, and we say the ratio $a{:}b$ is a root of R. In this sense it is true that $R(x_0, x_1)$ has at most n roots. Over an algebraically closed field, $R(x_0, x_1)$ can be completely factored into exactly n factors of the form $b_ix_0 - a_ix_1$ corresponding to the ratio roots $a_i{:}b_i$. If some factor occurs k times, then the corresponding ratio is called a root of multiplicity k. In this sense then, $R(x_0, x_1)$ has exactly n roots over an algebraically closed field.

9. IDEALS

An *ideal* in the integral domain $F[\mathbf{x}]$ of all polynomials in n variables over a field F is a nonempty subset $\mathfrak{I}$ of $F[\mathbf{x}]$ satisfying the following two conditions. If $p \in \mathfrak{I}$ and $q \in \mathfrak{I}$, then $p + q \in \mathfrak{I}$; if $p \in \mathfrak{I}$ and $f \in F[\mathbf{x}]$, then

$pf \in \mathfrak{J}$. The *sum* of two subsets $\mathfrak{A}$, $\mathfrak{B}$ of $F[\mathbf{x}]$ is $\mathfrak{A} + \mathfrak{B} = \{a + b \mid a \in \mathfrak{A}$ and $b \in \mathfrak{B}\}$; the *product* is $\mathfrak{A}\mathfrak{B} = \{\sum_{i=1}^{n} a_i b_i \mid \text{all } a_i \in \mathfrak{A}, \text{ all } b_i \in \mathfrak{B}, n = 1, 2, \cdots\}$. The sum and the product of two ideals are ideals. If $\mathfrak{A}$ consists of one element f, we write $f\mathfrak{B}$ for $\mathfrak{A}\mathfrak{B}$. We have $\mathfrak{A}\mathfrak{B} = \mathfrak{B}\mathfrak{A}$, $\mathfrak{A} + \mathfrak{B} = \mathfrak{B} + \mathfrak{A}$, $\mathfrak{A}(\mathfrak{B}\mathfrak{C}) = (\mathfrak{A}\mathfrak{B})\mathfrak{C}$, and $\mathfrak{A}(\mathfrak{B} + \mathfrak{C}) = \mathfrak{A}\mathfrak{B} + \mathfrak{A}\mathfrak{C}$ for all subsets $\mathfrak{A}$, $\mathfrak{B}$, $\mathfrak{C}$ of $F[\mathbf{x}]$.

An ideal $\mathfrak{J}$ in $F[\mathbf{x}]$ is called *principal* if $\mathfrak{J} = fF[\mathbf{x}]$ for some f in $F[\mathbf{x}]$. $\mathfrak{J}$ is called *prime* if, for all f and g in $F[\mathbf{x}]$, $fg \in \mathfrak{J} \rightarrow f \in \mathfrak{J}$ or $g \in \mathfrak{J}$.

We shall need the following version of the *Hilbert basis theorem:* If $\mathfrak{J}_1, \mathfrak{J}_2, \cdots$ are ideals in $F[\mathbf{x}]$ such that $\mathfrak{J}_1 \subset \mathfrak{J}_2 \subset \cdots$, then there exists an integer N such that $\mathfrak{J}_N = \mathfrak{J}_{N+1} = \mathfrak{J}_{N+2} = \cdots$.

10. NUMBER THEORY

We are concerned here only with the ordinary (rational) integers.

If n and m are integers, we say n *divides* m, written $n|m$, if $m = qn$ for some integer q. A *prime* is an integer $p > 1$ whose only divisors are ± 1 and $\pm p$. If p is a prime and $p|mn$, then $p|m$ or $p|n$. This leads to the fact that every integer $n > 1$ has a unique prime decomposition

$$n = p_1^{e_1} p_2^{e_2} \cdots p_k^{e_k}, \tag{*}$$

where the e_i are positive integers and $p_1 < p_2 < \cdots < p_k$ are primes.

Let a, b, and m be integers. We say a is *congruent to b modulo m*, written $a \equiv b \pmod{m}$, if $m \mid (a - b)$. Congruence mod m is an equivalence relation satisfying, for all a, b, c, $a \equiv b \pmod{m} \rightarrow a + c \equiv b + c \pmod{m}$ and $ac \equiv bc \pmod{m}$.

Two integers are *relatively prime* if they have no common prime divisors. m and n are relatively prime iff there exist integers a and b such that $am + bn = 1$.

An integer n is a *square* if $n = m^2$ for some integer m. Every nonnegative integer is the sum of four squares. The proof of this fact uses *Lagrange's identity:*

$$\begin{aligned}(a_1^2 + a_2^2 + a_3^2 + a_4^2)(b_1^2 + b_2^2 + b_3^2 + b_4^2) = \\ (a_1b_1 + a_2b_2 + a_3b_3 + a_4b_4)^2 \\ + (a_1b_2 - a_2b_1 + a_3b_4 - a_4b_3)^2 \\ + (a_1b_3 - a_3b_1 + a_2b_4 - a_4b_2)^2 \\ + (a_1b_4 - a_4b_1 + a_2b_3 - a_3b_2)^2.\end{aligned}$$

An integer $n > 1$ is the sum of two squares iff in its prime decomposition (*) we have $2|e_i$ whenever $p_i \equiv 3 \pmod{4}$. An immediate consequence of this is the following corollary.

COROLLARY 1 If m and n are each sums of two squares and if $m = kn$, then k is the sum of two squares.

We shall now prove two special results that are corollaries of the following famous theorem of Euler: If m and n are relatively prime positive integers, then $m^{\phi(n)} \equiv 1 \pmod{n}$, where $\phi(n)$ is the number of integers between 1 and n inclusive that are relatively prime to n.

COROLLARY 2 If m and j are positive integers and p is a prime, then $m^{p^j} \equiv m^{p^{j-1}} \pmod{p^j}$.

PROOF. It is easy to see that $\phi(p^j) = p^j - p^{j-1}$. If p and m are relatively prime, then so are p^j and m, and then, by the theorem of Euler, $m^{p^j - p^{j-1}} \equiv 1 \pmod{p^j}$. This implies the required congruence. If p and m are not relatively prime, then $p|m$. But then p^j divides both sides of the required congruence, so it still holds. ▎

COROLLARY 3 If p is a prime, n a positive integer not divisible by p, and k any positive integer, then there is an integer x such that $xn \equiv 1 \pmod{p^k}$.

PROOF. p and n are relatively prime, hence p^k and n are relatively prime. By the theorem of Euler there is a positive integer m such that $n^m \equiv 1 \pmod{p^k}$. Take x to be n^{m-1}. ▎ (A generalization of this corollary follows easily from the remarks in the paragraph before Lagrange's identity.)

11. GROUPS

For a definition of group see Section 3.

The *inverse* of an element g of a group $(G, \cdot)$ is the unique element g^{-1} of G such that $gg^{-1} = 1$, the identity element. We have $(ab)^{-1} = b^{-1}a^{-1}$ for all a and b in G.

A *subgroup* of a group $(G, \cdot)$ is a subset of G, which is a group under the restricted operation of G. A nonempty subset H of G is a subgroup of G iff for all x and y in H we have $xy^{-1} \in H$. If H is a subgroup of a finite group G, then by *Lagrange's theorem*, $|H|$ divides $|G|$.

The *order* of an element x of a finite group G is the smallest positive integer n such that $x^n = 1$. The order of any element of G divides $|G|$. If G contains more than one element, then it contains an element of prime order.

If A and B are subsets of a group G, we define AB to be $\{ab \mid a \in A \text{ and } b \in B\}$. If A consists of a single element a, we write aB for AB. For all subsets A, B and C, we have $A(BC) = (AB)C$. If H and K are finite subgroups of G, then $|HK|\,|H \cap K| = |H|\,|K|$.

A *normal subgroup* of a group G is a subgroup H such that, for all g in G, $gH = Hg$, that is, $gHg^{-1} = H$.

The *normalizer* of a subset A of a group G is $N_G(A) = \{x \in G \mid xA = Ax\}$. $N_G(A)$ is a subgroup of G. Hence, if G is finite, $|G|/|N_G(A)|$ is an integer. This integer equals the number of distinct subsets of G of the form gAg^{-1} for some g in G. These subsets, called the *conjugates* of A, each have the same number of elements as A.

The intersection of any family of subgroups of a group G is a subgroup of G. If A is a subset of a group G, the group $G(A)$ *generated* by A is the intersection of the family of all subgroups of G that contain A. If A consists of a single element g of finite order n, then $G(A) = \{1, g, g^2, \cdots, g^{n-1}\}$.

Let p be a prime. A group G is called a *p-group* if $|G| = p^n$ for some positive integer n. We need the following two basic results from the theory of p-groups.

1. A finite group G is a p-group iff $|G| > 1$ and every element of G has order a power of p.

2. If G is a p-group, then G contains a nonidentity element g such that $gx = xg$ for all x in G.

12. PERMUTATION GROUPS

A *permutation* of a set is a one–one correspondence between the set and itself. If f is a permutation of the finite set $\{x_1, x_2, \cdots, x_n\}$, we write

$$f = \begin{pmatrix} x_1 & x_2 & \cdots & x_n \\ f(x_1) & f(x_2) & \cdots & f(x_n) \end{pmatrix}.$$

The set of all permutations of a set S forms a group under function composition. Any subgroup of this group is called a *permutation group on S*.

A *cyclic permutation* of the ordered n-tuple $(x_1, x_2, \cdots, x_n)$ is a permutation of $\{x_1, \cdots, x_n\}$ that is of the form

$$\begin{pmatrix} x_1 & x_2 & \cdots & x_{n-k+1} & x_{n-k+2} & \cdots & x_n \\ x_k & x_{k+1} & & x_n & x_1 & & x_{k-1} \end{pmatrix}$$

for some k. This permutation is denoted by $(x_1x_2 \cdots x_n)^k$. The set of all cyclic permutations on a given finite set forms a permutation group on that set.

Let G be a permutation group on a set S. A subset T of S is a *G-class* if the following two conditions hold: given any x and y in T, there is a g in G such that $g(x) = y$; if $x \in T$ and $g \in G$, then $g(x) \in T$. The G-classes form a partition of S. G is *transitive* on S if S is a G-class. In other words, G is transitive on S if, given any ordered pair (x, y) of elements of S, there is a g in G such that $g(x) = y$. G is *sharply transitive* on S if this g is uniquely determined by the ordered pair (x, y). A permutation g in G is said to *fix* an element x of S if $g(x) = x$. G (transitive) is sharply transitive on S iff the only permutation in G that fixes any element of S is the identity of G.

G is *doubly transitive* if, given any pairs (a, b), (a', b') of distinct elements, there is a permutation in G that sends a to a' and b to b'.

Let G be a permutation group on S. Let T be a G-class. Let H be a subgroup of G. Then every H-class is either contained in or disjoint from T. Therefore, the H-classes that are contained in T partition T.

Finally, we prove two special lemmas on permutation groups, which will be needed.

LEMMA 1 Let G be a finite permutation group on a finite set S. Suppose there is a prime p such that for each x in S there is a g in G satisfying: (i) g is of order p; (ii) g fixes only x. Then G is transitive on S.

PROOF. Suppose G is not transitive on S. Then there is more than one G-class. Let S_1, S_2 be two G-classes. Choose x in S_1 and let g satisfy (i) and (ii). Let $H = \{1, g, \cdots, g^{p-1}\}$ be the subgroup generated by G. $\{x\}$ is an H-class. Let T be any other H-class. Obviously, $|T| \leqq p$, since $|H| = p$. Suppose $|T| < p$. Then for some $y \in T$ and integers i, j with $0 < i < j < p$, we have $g^i(y) = g^j(y)$. Then $g^m(y) = y$, where $m = j - i$. m and p are relatively prime, so according to Section 10 there exist integers a and b such that $am + bp = 1$. Then $g(y) = g^{am+bp}(y) = y$, since $g^p = 1$ and $g^m(y) = y$. But then, by (ii), $y = x$, a contradiction. Hence, $|T| = p$. Thus, S_1 is partitioned into $\{x\}$ and H-classes of order p and S_2 is partitioned just into H-classes of order p. Thus, p divides $|S_2|$ but not $|S_1|$. But, if we choose x in S_2 and repeat the argument, we have that p divides $|S_1|$ but not $|S_2|$, a contradiction. ∎

LEMMA 2 Let G be a permutation group on a set S. If G is transitive on S and G is Abelian, then G is sharply transitive on S.

PROOF. Suppose $g(x) = x$ for some s in S and g in G. We must show that g is the identity, that is, $g(y) = y$ for all y in S. Given any y in S there is a g' such that $g'(x) = y$. Then $g(y) = gg'(x) = g'g(x) = g'(x) = y$. ∎

II
Projective Spaces

1. THE REAL PROJECTIVE PLANE

This section contains intuitive motivation for the abstractions and generalizations to follow.

* * *

In Figure II.1.1 seven parallel equally spaced line segments are shown as being projected by means of a light source S from a glass plate **P** onto a screen **P′**. The image consists of parallel line segments on **P′**, no longer equally spaced. **P′** is long enough to catch the images of only the first five segments, but it could be extended to catch the image of the sixth. However, no matter how far **P′** is extended, it will not catch the image of the seventh segment; the light rays from S that pass through segment 7 are parallel to **P′**.

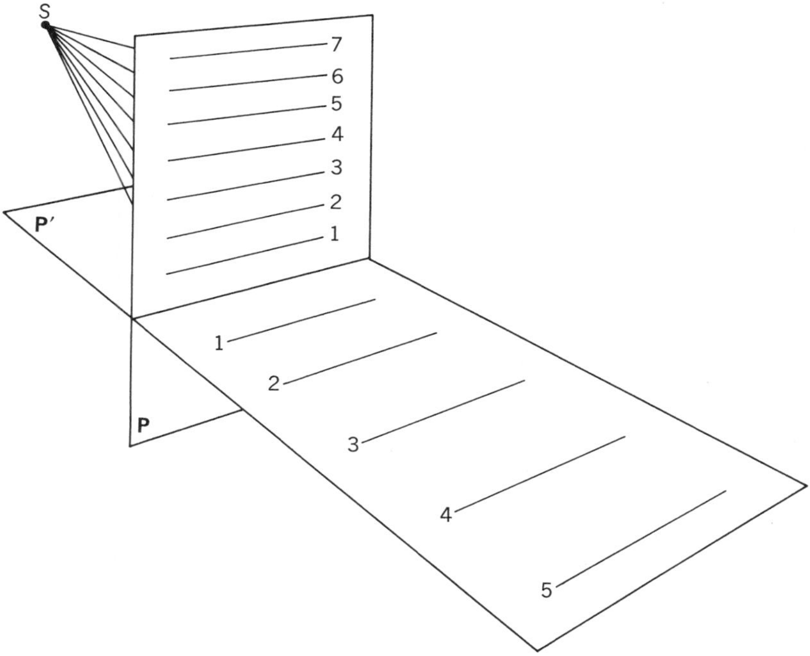

Fig. II.1.1

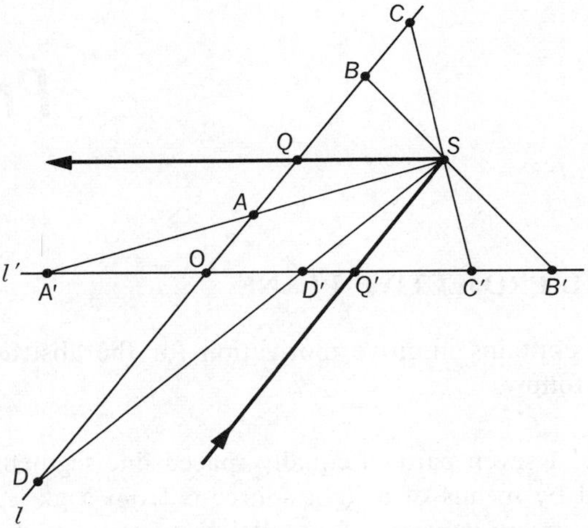

Fig. II.1.2

In some sense, segment 7 "has an image at infinity on **P**′ ". Our first goal is to make this idea more precise.

In order to simplify the situation, we will first consider the planar version of Figure II.1.1. Then, instead of planes **P** and **P**′ we have lines, say, l and l' (Figure II.1.2), and a source point S on neither line. *Projection* from l to l' with *center* S is defined as follows: The image of a point P on l is the point P' on l' collinear with P and S. In other words, $P' = PS \wedge l'$, where PS is the line joining P and S and $PS \wedge l'$ is the point of intersection of PS and l'. In Figure II.1.2, for example, A is projected to A' and B to B', although in the case of B we are extending the physical idea of projection. Point $O = l \wedge l'$ is projected onto itself.

The problem encountered in Figure II.1.1 is illustrated in the planar case (Figure II.1.2) by point Q, which is not projected to any point of l'. There is another problem in this figure—point Q' is not the projected image of any point on l. As these problems obviously stem from the existence of parallel lines, we shall solve them by causing all lines to intersect! More precisely, we shall embed the Euclidean plane in a larger geometrical structure in which each pair of lines has a common point.

Imagine that each class of parallel lines passes through a new point adjoined to the plane. A good way to picture this situation is to imagine the entire Euclidean plane $\mathbf{E}_2$ to be inside the circle of Figure II.1.3. Pick a point O in $\mathbf{E}_2$. Add one new point, outside $\mathbf{E}_2$, to each line through O. If l is a line of $\mathbf{E}_2$ not through O, there is a unique line l' through O parallel to l. Put the new point of l' also on l. Finally, imagine one new line that contains all of the new points and no others. The resulting extension of $\mathbf{E}_2$

is called the *real projective plane* and is denoted $\mathbf{P}_2\mathrm{R}$. We must mention the real field R and dimension 2 in the notation because of generalizations to come.

The reader should check that in $\mathbf{P}_2\mathrm{R}$ each pair of distinct points is on a unique line and, as was our goal, that each pair of distinct lines has a unique common point. The planar version of our problem is solved: In Figure II.1.2, Q is projected to the new point of l' and the new point of l is projected to Q'.

Returning to the original problem of Figure II.1.1, we must embed Euclidean space $\mathbf{E}_3$ in a geometrical structure in which every line intersects every plane. The procedure is essentially the three-dimensional analog of the above. Pick a point O in $\mathbf{E}_3$. To each plane $\mathbf{P}$ through O add a new line and require it to be on every plane parallel to $\mathbf{P}$. To each old line l add a new point and require it to be on the new line of every plane that contains l. Finally, form a new plane of all the new points and lines. We have constructed *real projective space* $\mathbf{P}_3\mathrm{R}$. The reader should check that every line meets every plane in $\mathbf{P}_3\mathrm{R}$ and that two distinct planes always meet in a line. In this context "line" means either "old line with new point adjoined" or "new line," and "plane" means either "old plane with new line adjoined" or "new plane."

By *projection* from a plane $\mathbf{P}$ to a plane $\mathbf{P}'$ with *center* a point S not on $\mathbf{P}$ or $\mathbf{P}'$, we mean the function sending a point P on $\mathbf{P}$ to the point $PS \wedge \mathbf{P}'$, where PS meets $\mathbf{P}'$. In view of the above remarks, then, the space version of our problem is solved.

Having constructed geometries $\mathbf{P}_2\mathrm{R}$ and $\mathbf{P}_3\mathrm{R}$ in which projection is always possible, we naturally inquire into the effects of projection on various objects and concepts in the geometry. For example, under projection in $\mathbf{P}_3\mathrm{R}$ from

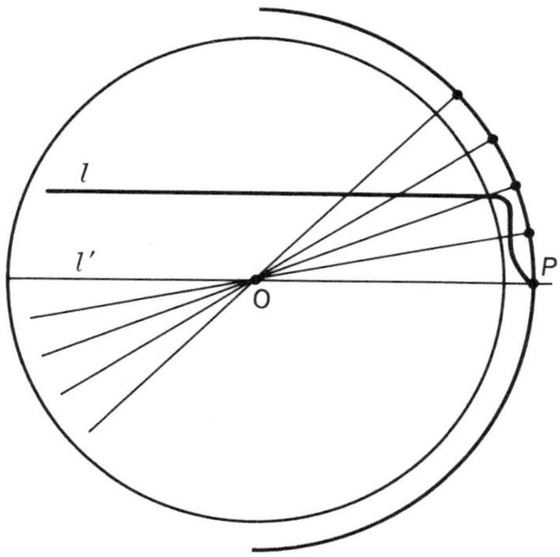

Fig. II.1.3

one plane to another, the image of a point is always a point and the image of a line is always a line. Thus, these objects are worthy of study in projective geometry. On the other hand, the image of a circle need not be a circle; hence, these objects cannot be studied in projective geometry. (It may be intuitively clear, however, that under projection in $\mathbf{P}_3\mathrm{R}$ the image of a conic is always a conic.) The projection in Figure II.1.2 sending A, B, C to A', B', C', respectively, suggests that in $\mathbf{P}_2\mathrm{R}$ the concept of a point being between two others on a line is not preserved. However, the weaker concept of two points "separating" two others on a line is preserved. Thus, in Figure II.1.2 A and C separate B and D, and A' and C' separate B' and D'. We shall discuss separation in a more rigorous way in Section IV.6. For more on these and many other ideas in synthetic projective geometry, see [5] or [6].

This concludes our intuitive introduction to the classical projective geometries $\mathbf{P}_2\mathrm{R}$ and $\mathbf{P}_3\mathrm{R}$. We now have enough knowledge of $\mathbf{P}_2\mathrm{R}$ to allow us to introduce a system of coordinates. This will be done in the next section. After that, we shall proceed in the usual rigorous axiomatic style of mathematics.

Exercises

1. In $\mathbf{P}_3\mathrm{R}$ prove that: (i) each pair of distinct points is on a unique line; (ii) given two distinct points in a plane, the line through them is in that plane; (iii) each set of three noncollinear points is on a unique plane; (iv) each point-line pair, with the point not on the line, is on a unique plane; (v) two distinct lines are on a plane iff they intersect.
2. Draw a figure like Figure II.1.1 but omit segments 1–6 and make segment 7 a longer segment, s. Draw the image in $\mathbf{P}'$ of the following figures in $\mathbf{P}$: (i) two lines that intersect in a point below s; (ii) two lines that intersect in a point on s; (iii) a circle below s; (iv) a circle tangent to s; (v) a circle that is cut twice by s.
3. In $\mathbf{P}_2\mathrm{R}$, let A, B, C be distinct points on a line l, and A', B', C' distinct points on a line l'. (We may have $l = l'$.) Prove that there exist points S_1, S_2, S_3 and lines l_1, l_2 such that projection from l to l_1 with center S_1, then from l_1 to l_2 with center S_2, and then from l_2 to l' with center S_3 has the net effect of sending A to A', B to B', and C to C'. (We shall later prove a certain uniqueness statement.)

Theorems in Euclidean plane geometry that are concerned only with points, lines, and their incidence or nonincidence can often be proved easily as follows: Make a drawing such as that described in Exercise 2. Draw the figure for the theorem in plane $\mathbf{P}$ in such a way that the line giving the most trouble in the proof lies on s. Now prove the theorem for that part of the image of the figure in $\mathbf{P}'$ not on the new line, by using any previously known results of "ordinary" Euclidean geometry. We shall illustrate this technique in the next exercise, with its generous hint, and then leave two more exercises entirely to the reader.

4. Let l, m, n be distinct lines concurrent at a point P. Let A, A' be distinct points on l; B, B' distinct points on m; C, C' distinct points on n. Suppose that the points $D = AB \wedge A'B'$, $E = AC \wedge A'C'$, $F = BC \wedge B'C'$ exist. Then D, E, F are collinear. (Hint: Place DE on s. In $\mathbf{P}'$ use the theorem that two lines cut proportional segments on two intersecting lines iff they are parallel.)
5. Let l and m be distinct lines. Let A, B, C be distinct points on l; A', B', C' distinct points on l'. Suppose that the points $D = AB' \wedge A'B$, $E = AC' \wedge A'C$, and $F = BC' \wedge B'C$ exist. Then D, E, F are collinear.
6. Let A, B, C be distinct points on a line l. Let D, E be points not on l. Let m and n be lines through C that are distinct from l and do not pass through D or E. Suppose that the points $F = n \wedge AE$, $G = n \wedge BE$, $H = m \wedge BD$, and $I = m \wedge AD$ exist, and that the points $J = IB \wedge AH$ and $K = AG \wedge BF$ exist. Make a conjecture about $DJ \wedge EK$, supposing that it exists, and prove it.

2. COORDINATIZATION OF THE REAL PROJECTIVE PLANE

To coordinatize $\mathbf{P}_2\mathrm{R}$ we shall extend a coordinatization of $\mathbf{E}_2$. Suppose $\mathbf{E}_2$ has an ordinary Cartesian coordinate system. Extend $\mathbf{E}_2$ to $\mathbf{P}_2\mathrm{R}$ (Figure II.1.3). Each new point P corresponds to the line OP, where O is the origin of the coordinate system in $\mathbf{E}_2$. Distinct new points correspond in this way to lines having only O in common. If P and Q are distinct new points, it follows that the sets $C_P = \{(x, y) \mid (x, y) \text{ is on } OP \text{ and } (x, y) \neq (0, 0)\}$ and $C_Q = \{(x, y) \mid (x, y) \text{ is on } OQ \text{ and } (x, y) \neq (0, 0)\}$ are disjoint. Moreover, these sets are easy to describe. If (a, b) is any element of C_P then

$$C_P = \{t(a, b) \mid t \neq 0\}. \tag{1}$$

We could use these sets as coordinates of the new points. But, since new and ordinary points can be interchanged by projection (Figure II.1.2), they ought to have the same kind of coordinates. There is no geometrically meaningful way to "simplify" the sets C_P, since projections do not preserve distance. We shall therefore use sets of ordered *triples* of real numbers to achieve the desired uniformity of description.

Assign as the coordinate of a new point P the set

$$\{(x, y, 0) \mid (x, y) \text{ is on } OP \text{ and } (x, y) \neq (0, 0)\}. \tag{2}$$

By (1) this set is the same as the set $\{t(a, b, 0) \mid t \neq 0\}$, where $(a, b, 0)$ is any element of (2). Assign as the coordinate of an ordinary point (c, d) the set

$$\{t(c, d, 1) \mid t \neq 0\}. \tag{3}$$

The reader should check that any two distinct points of $\mathbf{P}_2\mathrm{R}$ have disjoint coordinate sets (Exercise 1). Note that a point with coordinate $\{t(a, b, c) \mid t \neq 0\}$ is new if and only if $c = 0$.

We shall connect this coordinatization scheme with the geometrical structure of $\mathbf{P}_2\mathrm{R}$ by introducing a system of coordinates for lines. In the Cartesian coordinate system of $\mathbf{E}_2$, a line consists of all points (x, y) satisfying an equation of the form $px + qy + r = 0$, where p, q are not both zero. Actually, each member of the family of equations

$$(kp)x + (kq)y + (kr) = 0, \quad \text{with } k \neq 0, \tag{4}$$

determines the line. Conversely, each line of $\mathbf{E}_2$ determines a family (4), with p, q not both zero. Suppose now that l is a line of $\mathbf{P}_2\mathrm{R}$, not the new line. Then the ordinary part of l has a family of equations of the form (4). Assign as coordinate for l in $\mathbf{P}_2\mathrm{R}$ the set

$$\{k(p, q, r) \mid k \neq 0\}. \tag{5}$$

The only set of the form (5) not used to coordinatize some line is the set

$$\{(0, 0, k) \mid k \neq 0\}. \tag{6}$$

We assign this set as the coordinate of the new line l_∞.

Now let P be an ordinary point with coordinate $\{t(a, b, 1) \mid t \neq 0\}$, and let l be an ordinary line with coordinate $\{k(p, q, r) \mid k \neq 0\}$. Consider the dot product $k(p, q, r) \cdot t(a, b, 1) = kt(pa + qb + r)$. Since $kt \neq 0$, this dot product is 0 iff $pa + qb + r = 0$, that is, P is on l. In the next two paragraphs we show that this result (that the dot product is zero iff the point is on the line) holds for *all* points and lines in $\mathbf{P}_2\mathrm{R}$. That will establish the desired connection between the coordinatization scheme and the geometrical structure of $\mathbf{P}_2\mathrm{R}$.

A point P with coordinate $\{t(a, b, c) \mid t \neq 0\}$ is on l_∞, with coordinate (6), if and only if P is new, that is, if and only if $c = 0$. But for any nonzero t and k, $c = 0$ is equivalent to $0 = ktc = (0, 0, k) \cdot t(a, b, c)$, as desired.

Let P be a new point with coordinate $\{t(a, b, 0) \mid t \neq 0\}$, and l an ordinary line with coordinate $\{k(p, q, r) \mid k \neq 0\}$. We wish to show that P is on l if and only if $k(p, q, r) \cdot t(a, b, 0) = 0$, that is, if and only if

$$ap + bq = 0. \tag{7}$$

Suppose P is on l. Then P is on the line l' through O parallel in $\mathbf{E}_2$ to l. A Cartesian coordinate equation for l is $px + qy + r = 0$; hence, an equation for l' is $px + qy = 0$. The point $(q, -p) \neq (0, 0)$ is clearly on l'. Thus, by definition the new point on l', namely, P, has coordinate $\{t(q, -p, 0) \mid t \neq 0\}$. This set must coincide with the original coordinate $\{t(a, b, 0) \mid t \neq 0\}$ of P. Hence, $(a, b, 0) = \bar{t}(q, -p, 0)$ for some $\bar{t}$. Then $pa + qb = \bar{t}(qp - pq) = 0$, so that (7) holds. Conversely, suppose that (7) holds. If $p \neq 0$, then $a = -(bp^{-1})q$, from which we see that $(a, b, 0) = (-bp^{-1})(q, -p, 0)$. Then $\{t(a, b, 0) \mid t \neq 0\} = \{t(q, -p, 0) \mid t \neq 0\}$, so P is the new point on l'. Hence, P is on l. If $p = 0$, then $q \neq 0$, since l' is ordinary, and so from (7), $b = 0$. Then $(a, b, 0) = (a, 0, 0) = (aq^{-1})(q, -p, 0)$, and, as before, P is on l.

We summarize these results as follows: Two triples $\mathbf{a}$, $\mathbf{b}$ of real numbers are said to be *proportional* if $\mathbf{a} = x\mathbf{b}$ for some x. Proportionality of nonzero elements of R^3 is an equivalence relation. The corresponding equivalence classes are the coordinate sets above. If we identify points and lines with their coordinate sets, we have an analytic definition.

ANALYTIC DEFINITION OF $\mathbf{P}_2$R Points are the equivalence classes under proportionality of nonzero elements of R^3; so are lines. A point P is on a line l if and only if $\mathbf{p} \cdot \mathbf{l} = 0$ for some (hence, all) $\mathbf{p}$ in P and $\mathbf{l}$ in l.

If A is a geometric object and α a symbol, we abbreviate the sentence "A has the coordinate α" by writing $A \vdots \alpha$. The coordinate system introduced above is the *homogeneous* (projective) coordinatization of $\mathbf{P}_2$R. If a point $P \vdots \{t(a, b, c) \mid t \neq 0\}$ in this system, we will allow ourselves to write $P \vdots (ta, tb, tc)$ for any particular $t \neq 0$; if a line $l \vdots \{k(p, q, r) \mid k \neq 0\}$, we may write $l \vdots [kp, kq, kr]$. We will even speak of "the point (a, b, c)" or "the line $[p, q, r]$". We have seen there is no danger in this; however, care must be exercised whenever a new concept is introduced (compare Section V.1).

Analytic geometry in $\mathbf{P}_2$R is merely applied 3×3 linear algebra. For example, let us find the line $[p, q, r]$ joining points $(1, 2, 1)$ and $(2, 1, 2)$. We first note that these are distinct points. (Points $(1, 2, 1)$ and $(\frac{1}{2}, 1, \frac{1}{2})$, for example, are not distinct.) Now we find any nonzero simultaneous solution of the equations

$$\begin{aligned} p + 2q + r &= 0 \\ 2p + q + 2r &= 0, \end{aligned}$$

say, $p = 1, q = 0, r = -1$. The line is $[1, 0, -1]$, or equivalently, $[k, 0, -k]$ for any $k \neq 0$. As another example we prove a useful theorem.

THEOREM 1 Let P, Q be distinct points in $\mathbf{P}_2$R. A point X is on line PQ if and only if $X \vdots a\mathbf{p} + b\mathbf{q}$ for some a, b, not both zero, where $P \vdots \mathbf{p}$, $Q \vdots \mathbf{q}$. If $X \neq Q$ we may set $a = 1$.

PROOF. Suppose $PQ \vdots \mathbf{u}$. If $X \vdots a\mathbf{p} + b\mathbf{q}$, then $(a\mathbf{p} + b\mathbf{q}) \cdot \mathbf{u} = a(\mathbf{p} \cdot \mathbf{u}) + b(\mathbf{q} \cdot \mathbf{u}) = 0$, so X is on PQ. Conversely, if $X \vdots \mathbf{x}$ is on PQ, then $\mathbf{x} \cdot \mathbf{u} = 0$, so $\mathbf{x}$ is in the orthogonal complement U of the one-dimensional vector space with basis $\{\mathbf{u}\}$ (Section I.4). U is of dimension $3 - 1 = 2$. $\mathbf{p}$ and $\mathbf{q}$ are in U and, since $P \neq Q$, $\mathbf{p}$ and $\mathbf{q}$ are independent. Therefore, U is spanned by $\mathbf{p}$ and $\mathbf{q}$. In particular, $\mathbf{x} \in U$ is a linear combination $a\mathbf{p} + b\mathbf{q}$ of $\mathbf{p}$ and $\mathbf{q}$, that is, $X\vdots a\mathbf{p} + b\mathbf{q}$ for some a, b. If $X \neq Q$, then $a \neq 0$, and a coordinate for X is $a^{-1}(a\mathbf{p} + b\mathbf{q}) = \mathbf{p} + (a^{-1}b)\mathbf{q}$. ∎

Exercises

1. Prove that the family of all sets of the form (2) or (3) forms a partition of $\mathrm{R}^3 - \{(0, 0, 0)\}$.
2. In $\mathbf{P}_2\mathrm{R}$ suppose $A\vdots(1, 1, -1)$, $B\vdots(2, 0, 1)$, $C\vdots(1, 0, 2)$, $l\vdots[-1, -1, 2]$, $m\vdots[8, 2, 7]$. Find a coordinate for $(AB \wedge l)C \wedge m$. [An answer is $(7, 0, -8)$.]
3. Show how $\mathbf{P}_2\mathrm{R}$ can be represented by the set of all lines in $\mathbf{E}_3$ through the origin, or what amounts to the same thing, by the set of all (unordered) pairs of diametrically opposite points on the unit sphere in $\mathbf{E}_3$.
4. Let A be a 3×3 matrix of real numbers. Consider the following function on $\mathbf{P}_2\mathrm{R}$:

$$\text{Point}\quad \{t(x, y, z)\} \to \{t(x, y, z)A\};$$

$$\text{Line}\quad \{t[x, y, z]\} \to \left\{tA\begin{pmatrix} x \\ y \\ z \end{pmatrix}\right\}.$$

(a) If A is nonsingular, prove that this function is one–one and onto on points and lines, and that it preserves incidence.
(b) What can be said if A is singular?
5. In $\mathbf{P}_2\mathrm{R}$ suppose that $A\vdots(a_0, a_1, a_2)$, $B\vdots(b_0, b_1, b_2)$, $A \neq B$. Prove that

$$AB\vdots\left[\begin{vmatrix} a_1 & b_1 \\ a_2 & b_2 \end{vmatrix}, \begin{vmatrix} a_2 & b_2 \\ a_0 & b_0 \end{vmatrix}, \begin{vmatrix} a_0 & b_0 \\ a_1 & b_1 \end{vmatrix}\right].$$

(The entries are determinants.)
6. Let A, B, C, D be four points of $\mathbf{P}_2\mathrm{R}$, no three collinear. Prove that there exist coordinate vectors $\mathbf{a}$, $\mathbf{b}$, $\mathbf{c}$, $\mathbf{d}$, of A, B, C, D, respectively, such that $\mathbf{d} = \mathbf{a} + \mathbf{b} + \mathbf{c}$.
7. Let $\{A_n\}$ be a sequence of points and A a point of $\mathbf{P}_2\mathrm{R}$. By $\lim A_n = A$ we mean that there exist coordinate vectors, $A_n\vdots\mathbf{a}_n$ and $A\vdots\mathbf{a}$, such that $\lim_{n\to\infty} \mathbf{a}_n = \mathbf{a}$ in the usual sense. (See, for instance, [1].) Show that no sequence of points can have two distinct points as limits.
8. (See Exercise 7.) Prove that $\mathbf{P}_2\mathrm{R}$ is *compact*, in the sense that every

sequence of points has a convergent subsequence. (Use the Bolzano–Weierstrass theorem.)

9. (See Exercise 7.) A sequence $\{l_n\}$ of lines of $\mathbf{P}_2R$ is said to converge to a line l if there are coordinate vectors $l_n\!:\!\mathbf{l}_n$, $l\!:\!\mathbf{l}$ such that $\lim_{n\to\infty} \mathbf{l}_n = \mathbf{l}$. Suppose $\{A_n\}$ and $\{B_n\}$ are sequences of points that converge to A and B, respectively. Suppose further that $A_n \neq B_n$ for all n and that $A \neq B$. Prove $\lim_{n\to\infty} A_nB_n = AB$.
10. (For topology students.) $\mathbf{P}_2R$ is topologically equivalent to the rectangle shown in Figure II.2.1 with points on the opposite sides of the boundary

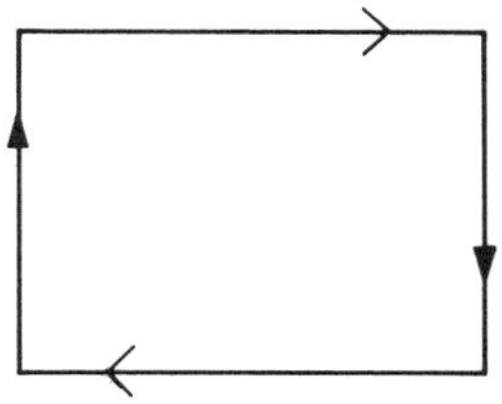

Fig. II.2.1

identified as shown (see Exercise 3). Show how to divide $\mathbf{P}_2R$ into ten triangles each, two of which have a common edge or have one common vertex or are disjoint. (This is called a *triangulation* of $\mathbf{P}_2R$.)

3. SYNTHETIC PROJECTIVE SPACES

Using a few of the most basic properties of $\mathbf{P}_2R$ and $\mathbf{P}_3R$, we shall now define a very general abstract geometry. The study of this interesting geometrical structure is one of the principal concerns of this book.

* * *

The fundamental properties of incidence in the real projective plane are simple: each pair of distinct points is on exactly one line; each pair of distinct lines is on exactly one point. The second of these two properties is not true in real projective space. We must say that each pair of distinct lines *in the same plane* are on exactly one point. This can be most effectively stated in the following form, known as the *Veblen–Young* axiom.*

If A, B, C, D are four points, no three collinear, and if AB intersects CD, then AC intersects BD.

---

* This is a slight simplification of Axiom A3 in O. Veblen and J. W. Young's famous two-volume work *Projective Geometry* (1910). The basic idea goes back to Pasch (see [11], Axiom II, 5).

Using this axiom, we shall base the theory of abstract projective geometry on three undefined terms: *point*, *line*, and *incidence*. Terms such as *plane* will be defined using these basic terms.

Consider an ordered triple $(\mathcal{P}, \mathcal{L}, \epsilon)$ where $\mathcal{P}$ and $\mathcal{L}$ are disjoint sets and ϵ is a relation from $\mathcal{P}$ to $\mathcal{L}$ (that is, a subset of $\mathcal{P} \times \mathcal{L}$). Elements of $\mathcal{P}$ are called *points;* elements of $\mathcal{L}$ are called *lines;* ϵ is called the *incidence relation*. If, for a point P and line l, (P, l) is in the relation ϵ, we write $P \epsilon l$, or sometimes $l \ni P$. This may be read "P is on l," "l passes through P," and so forth. If a point P is not on a line l, we write $P \not\epsilon l$. If several points $P, Q, \cdots$ all lie on the same line, we say that $P, Q, \cdots$ are *collinear*. If several lines $l, m, \cdots$ all pass through the same point, we say that $l, m, \cdots$ are *concurrent*. Indeed, we may use any familiar geometric terminology or notation whose meaning depends only on the terms point, line, and incidence.

DEFINITION 1 A *projective space* is a triple $(\mathcal{P}, \mathcal{L}, \epsilon)$, as described above, satisfying the following.

(1) Each two distinct points are on exactly one line.
(2) The Veblen–Young axiom holds.
(3) Every line contains at least three points.

Axiom (3) serves to rule out certain special cases (Exercise 2). A projective space $(\mathcal{P}, \mathcal{L}, \epsilon)$ is often denoted by a single boldface letter, such as **S**. We may write $P \in \mathbf{S}$ or $l \in \mathbf{S}$ instead of the more correct notation $P \in \mathcal{P}$ or $l \in \mathcal{L}$. We have used the epsilon both for the set-theoretic relation "is an element of" and for the arbitrary relation in $\mathcal{P} \times \mathcal{L}$. That this can cause no serious confusion is clear from the following theorem.

THEOREM 1 Let $(\mathcal{P}, \mathcal{L}, \epsilon)$ be a projective space. There is a correspondence between lines l in $\mathcal{L}$ and certain subsets $\bar{l}$ of $\mathcal{P}$ such that $P \epsilon l$ if and only if $P \in \bar{l}$.

PROOF. The correspondence is, obviously, $\bar{l} = \{P \in \mathcal{P} \mid P \epsilon l\}$. ▌
The set $\bar{l}$ defined above is called the *range* with *axis* l. Similarly, if P is a point the set $\tilde{P} = \{l \in \mathcal{L} \mid P \epsilon l\}$ is the *pencil* with *center* P.

In $\mathbf{P}_2$R the lines joining a point P to the points of a line l not through P cover all of $\mathbf{P}_2$R. This suggests a definition.

DEFINITION 2 Let P be a point, l a line in a projective space **S**, with P not on l. The *plane* $\mathbf{P}(P, l)$ determined by P and l is the ordered triple $(\mathcal{P}', \mathcal{L}', \epsilon')$ where

$\mathcal{P}' = \{\text{points } Q \in \mathbf{S} \mid Q = P \text{ or } PQ \text{ intersects } l\}$;
$\mathcal{L} = \{\text{lines } m \in \mathbf{S} \mid m \text{ passes through two distinct points of } \mathcal{P}'\}$;
ϵ' is the incidence relation of **S**, restricted to $\mathcal{P}'$ and $\mathcal{L}'$.

We shall also find it useful to have a direct abstraction of $\mathbf{P}_2$R.

DEFINITION 3 A *projective plane* is a triple $(\mathcal{P}, \mathcal{L}, \epsilon)$, as described preceding Definition 1, satisfying the following.

(a) Each pair of distinct points is on exactly one line.
(b) Each pair of distinct lines passes through at least one point.
(c) $\mathcal{P}$ contains four points, no three collinear.

As before, Axiom (c) serves merely to eliminate annoying special cases (Exercise 1). The next theorem is easy to conjecture but rather tedious to prove.

THEOREM 2 Every plane in a projective space is a projective plane.

PROOF. Let $\mathbf{P} = \mathbf{P}(P, l)$ be a plane in the projective space $\mathbf{S}$. Using (1), (2), (3) of Definition 1, we must show that $\mathbf{P}$ satisfies (a), (b), (c) of Definition 3.

(a) By (1) each pair of distinct points of $\mathbf{P}$ is on exactly one line; by Definition 2, this line is in $\mathbf{P}$.

(b) Let m and n be distinct lines in $\mathbf{P}$. By Definition 2, $m = EF$ and $n = GH$ for some E, F, G, H in $\mathbf{P}$. We may assume that these points are distinct. (Why?)

CASE 1. One of the lines, say m, is l. If $P \in n$, we may assume $P \neq G$. Then, since $G \in \mathbf{P}$, $PG = n$ intersects $l = m$ in a point which, by Definition 2, is in $\mathbf{P}$. Suppose $P \notin n$. Let $R = PG \wedge l$ and $S = PH \wedge l$. Then $GR \wedge SH = PR \wedge PH = P$, so by the Veblen–Young axiom, $RS \wedge GH = m \wedge n$ exists and, by Definition 2, is in $\mathbf{P}$.

CASE 2. Neither line is l. By Case 1, $M = m \wedge l$ and $N = n \wedge l$ exist in $\mathbf{P}$. Also by Case 1, $l \wedge EG = MN \wedge EG$ exists. Then by the Veblen–Young axiom, $Q = ME \wedge NG = m \wedge n$ exists. We must still show that $Q \in \mathbf{P}$. This is clear if $P = Q$ or $M = N$. Otherwise, note that $PM \in \mathbf{P}$. Then, since any two lines of $\mathbf{P}$ intersect (at least in $\mathbf{S}$), $PM \wedge n = PM \wedge QN$ exists. Hence, by the Veblen–Young axiom, $PQ \wedge MN = PQ \wedge l$ exists, and so, by Definition 2, $Q \in \mathbf{P}$.

(c) By (3) there are distinct points A, B on l and points C, D, on PA, PB, respectively, distinct from A, B, and P. By Definition 2, A, B, C, D are in $\mathbf{P}$. Collinearity of any three of these points would imply that $A = B$. Hence A, B, C, D, satisfy (c). ∎

The last arguments above have also established a corollary.

COROLLARY If a line l is in a plane $\mathbf{P}$ in a projective space, then every point on l is in $\mathbf{P}$.

We shall now study projective planes more closely. The following was implicit in the above proof.

THEOREM 3 Any two distinct lines in a projective plane intersect at exactly one point.

PROOF. Distinct lines have at least one common point, by (b) of Definition 3. If they had more than one, then (a) would be contradicted. ▎

THEOREM 4 Corresponding to each projective plane **P** there is a cardinal number n called the *order* of **P** such that:

(1) Each range of points in **P** is of cardinality $n + 1$.
(2) Each pencil of lines in **P** is of cardinality $n + 1$.
(3) The set of all points in **P** is of cardinality $n^2 + n + 1$.
(4) The set of all lines in **P** is of cardinality $n^2 + n + 1$.

PROOF. Let l, m be distinct lines in **P**. Let $n + 1$ be the cardinality of $\tilde{l}$, the range of all points on l. We first prove that there is a point P_0 that is on neither l nor m. Let P_1, P_2, P_3, P_4 be four points of **P**, no three of which are collinear. If some P_i is on neither l nor m, let it be P_0. Otherwise, we may assume $P_1 \in l$, $P_2 \in l$, $P_3 \in m$, $P_4 \in m$, all P_i distinct from $l \wedge m$. Then let $P_0 = P_1P_3 \wedge P_2P_4$. $P_0 \in l$ would imply $P_3 \in l$, a contradiction. Hence, $P_0 \notin l$. Similarly, $P_0 \notin m$.

Now define θ: $\tilde{l} \to \tilde{m}$ as follows: For $P \in l$ let $P\theta = P_0P \wedge m$. This defines a function on $\tilde{l}$, since $P \neq P_0$ and $P_0P \neq m$ for all $P \in l$. If $P\theta = Q\theta = T$, then $P = P_0T \wedge l = Q$. Hence, θ is one–one. If $T \in m$, then $S = P_0T \wedge l$ is on l and $S\theta = T$. Hence, θ is onto. Thus $\tilde{m}$ also has cardinality $n + 1$, and so (1) is proved.

Let $\tilde{P}$ be a pencil. Let l be a line not through P. Each line m of $\tilde{P}$ determines a unique point $m \wedge l$ of $\tilde{l}$ and each point Q of $\tilde{l}$ determines a unique line PQ of $\tilde{P}$. Hence, $\tilde{P}$ has the same cardinality as $\tilde{l}$, proving (2).

Let P be a point in **P**. Each of the $n + 1$ lines of $\tilde{P}$ contains exactly n points besides P. Every point $Q \neq P$ is on exactly one of these lines, namely, QP. Thus **P** contains $n(n + 1)$ points besides P. This proves (3).

Let l be a line in **P**. Each of the $n + 1$ points of $\tilde{l}$ contains exactly n lines besides l. Every line $m \neq l$ is on exactly one of these points, namely, $m \wedge l$. Thus **P** contains $n(n + 1)$ lines besides l. This proves (4). ▎

Of course, the order n of **P** may be a transfinite cardinal number. Students unfamiliar with the arithmetic of transfinite cardinals may wish to consult, for example, E. Kamke's *Theory of Sets*, Chapter II. However, it is sufficient for our purposes to remark that if some range of points in **P** is infinite then each two of sets (1), (2), (3), and (4), can be put in one–one correspondence.

The total similarity of the arguments for (3) and (4) strongly suggests the following very general result.

THEOREM 5 (The principle of duality in planes.) Given any theorem $\mathfrak{T}$ about points, lines, and incidence in a projective plane, the *dual* statement

$\mathfrak{T}^*$ obtained by interchanging the words "point" and "line" in the statement of $\mathfrak{T}$ is also a theorem.

PROOF. Let $\mathfrak{P}$ be a proof of $\mathfrak{T}$. Let $\mathfrak{P}^*$ be $\mathfrak{P}$ with the words "point" and "line" interchanged.

Appeals in $\mathfrak{P}$ to Axiom (a) of Definition 3 become in $\mathfrak{P}^*$ appeals to Theorem 3, and, hence, remain valid. Appeals in $\mathfrak{P}$ to (b) become in $\mathfrak{P}^*$ appeals to part of Axiom (a), and, hence, remain valid. Appeals in $\mathfrak{P}$ to (c) become appeals to (c*): **P** contains four lines l_1, l_2, l_3, l_4, no three concurrent. We show that (c*) is a consequence of Definition 3, and, hence, that $\mathfrak{P}^*$ is a valid proof of $\mathfrak{T}^*$. Let P_1, P_2, P_3, P_4 be four points, no three collinear. Then $l_1 = P_1P_2$, $l_2 = P_2P_3$, $l_3 = P_3P_4$, $l_4 = P_4P_1$ are four lines, no three on a common point. ∎

In dualizing an informally stated theorem or definition, we often must do more than just interchange "point" and "line." Thus, for example, "collinear points" dualizes to "concurrent lines," and "triangle" to "trilateral." Another approach that we use is to include the dual point and line concepts under one name. Thus, if "triangle" is interpreted as "three noncollinear points and the lines joining them," then the concept is *self-dual.* A little care must be exercised in applying Theorem 5 to a theorem of the form: "All projective planes that have property p have property s." If, as is often the case in practice, p is self-dual, then the dual theorem is: "All projective planes that have property p have property s^*." However, in *general,* the dual of the original theorem is: "All projective planes that have property p^* have property s^*."

As an example of the scope of Definition 3, in Figure II.3.1 we have a diagram of the smallest possible projective plane—often called the *Fano*

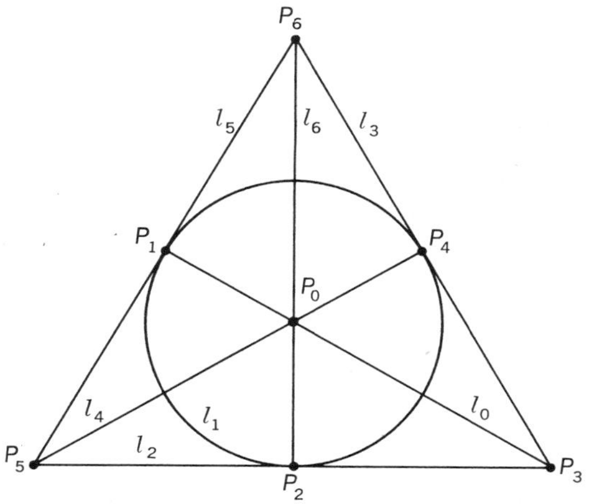

Fig. II.3.1

*configuration** (see Exercise 3). Note that points are shown by heavy dots. Not every intersection of the lines of the diagram is a point in the projective plane. It is usually convenient, though not necessary, to represent some of the lines of a plane by curved lines in a diagram. Thus one line of Figure II.3.1 is represented as a circle. This, of course, has no geometrical significance.

The idea in Definition 2 can be carried on by induction to define higher-dimensional spaces.

DEFINITION 4 A *two-dimensional projective subspace* of a projective space $\mathbf{S}$ is a plane in $\mathbf{S}$ (Definition 2). If $n > 2$ is an integer, if $\mathbf{P}_{n-1}$ is a previously defined $(n-1)$-dimensional subspace, and if P is a point of $\mathbf{S}$ not in $\mathbf{P}_{n-1}$, then the *n-dimensional projective subspace* $\mathbf{P}_n(P, \mathbf{P}_{n-1})$ is the ordered triple $(\mathcal{P}_n, \mathcal{L}_n, \epsilon_n)$, where: $\mathcal{P}_n = \{\text{points } Q \in \mathbf{S} \mid Q = P \text{ or } QP \text{ intersects } \mathbf{P}_{n-1}\}$; $\mathcal{L}_n = \{\text{lines } l \in \mathbf{S} \mid l \text{ passes through two distinct points of } \mathcal{P}_n\}$; ϵ_n is the incidence relation of $\mathbf{S}$ restricted to $\mathcal{P}_n$ and $\mathcal{L}_n$. With this we can define the dimension of a space.

DEFINITION 5 A projective space $\mathbf{S} = (\mathcal{P}, \mathcal{L}, \epsilon)$ has the following characteristics:

(a) Dimension -1 if $\mathcal{P} = \varnothing$.
(b) Dimension 0 if $\mathcal{P}$ consists of a single point.
(c) Dimension 1 if $\mathcal{L}$ consists of a single line.
(d) Dimension $n > 1$ if $\mathbf{S}$ contains a projective n-space $\mathbf{P}_n$ (Definition 4) such that $\mathbf{P}_n = \mathbf{S}$.
(e) Dimension ∞ if $\mathbf{S}$ contains a $\mathbf{P}_n$ for every n.

Part (e) of this definition could be considerably refined by means of transfinite numbers, but we shall not do this here. Our discussion of higher-dimensional spaces, with which we close this section, will be limited to the following few theorems and partial proofs. As we shall see at the end of the next chapter, this is not really a serious limitation.

THEOREM 6 Let $\mathbf{P}_2$ be a plane in a projective space. Let P be a point and l a line of $\mathbf{P}_2$, with P not on l. Then $\mathbf{P}_2 = \mathbf{P}_2(P, l)$.

PROOF. Exercise 6.

THEOREM 7 Let $\mathbf{P}_3$ be a three-dimensional projective subspace of a projective space. Let P be a point and $\mathbf{P}_2$ a plane of $\mathbf{P}_3$, with P not on $\mathbf{P}_2$. Then $\mathbf{P}_3 = \mathbf{P}_3(P, \mathbf{P}_2)$.

PROOF. Exercise 7.

---

* Gino Fano was the Italian mathematician who first considered finite projective geometries (1892).

COROLLARY (Compare with corollary to Theorem 2.) If a plane $\mathbf{P}_2$ is in a three-dimensional subspace $\mathbf{P}_3$ of a projective space, then every point and every line in $\mathbf{P}_2$ is in $\mathbf{P}_3$.

THEOREM 8 In a three-dimensional projective space, every line meets every plane.

PROOF. Let l be a line in $\mathbf{P}_3 = \mathbf{P}_3(P, \mathbf{P}_2)$. We first show that l meets $\mathbf{P}_2$. Say $l = AB$, where A and B are in $\mathbf{P}_3$. If A or B is in $\mathbf{P}_2$, then l meets $\mathbf{P}_2$. If $P \in l$ then, assuming $P \neq A$, $PA = l$ meets $\mathbf{P}_2$, since $A \in \mathbf{P}_3$. In the general case, PA and PB meet $\mathbf{P}_2$ in distinct points A' and B'. Then by the Veblen–Young axiom $AB = l$ meets $A'B'$. But $A'B'$ is in $\mathbf{P}_2$, so l meets $\mathbf{P}_2$. Finally, by Theorem 7, $\mathbf{P}_2$ could be any plane in $\mathbf{P}_3$. ▮

COROLLARY Two distinct planes in a projective 3-space intersect in exactly one line.

PROOF. Let the planes be $\mathbf{P}_2$ and $\mathbf{P}_2'$. Pick a point P in $\mathbf{P}_2$ not in $\mathbf{P}_2'$. Pick distinct lines l and m in $\mathbf{P}_2$ through P. Let $L = l \wedge \mathbf{P}_2'$, $M = m \wedge \mathbf{P}_2'$. Then $LM \subset \mathbf{P}_2 \wedge \mathbf{P}_2'$. If $Q \notin LM$ is also in $\mathbf{P}_2 \wedge \mathbf{P}_2'$, then $\mathbf{P}_2 = \mathbf{P}_2(Q, LM) = \mathbf{P}_2'$, a contradiction. Hence, $\mathbf{P}_2 \wedge \mathbf{P}_2' = LM$. ▮

Exercises

1. Find all types of systems $(\mathcal{P}, \mathcal{L}, \epsilon)$ that satisfy (a) and (b) but not (c) of Definition 3.
2. Find all types of systems $(\mathcal{P}, \mathcal{L}, \epsilon)$ that satisfy (1) and (2) but not (3) of Definition 1.
3. Show that every projective plane is of order at least two.
4. Let $\mathbf{S} = (\mathcal{P}, \mathcal{L}, \epsilon)$ be a projective space. Show that $\epsilon = \mathcal{P} \times \mathcal{L}$ if and only if $\dim \mathbf{S} \leq 1$.
5. Let $\mathbf{S}$ be a projective space. Prove that the following statements are equivalent: (1) $\mathbf{S}$ is a projective plane, (2) $\dim \mathbf{S} = 2$, (3) each pair of lines in $\mathbf{S}$ intersects, and $\mathbf{S}$ contains three noncollinear points.
6. Draw diagrams (such as Figure II.3.1) of projective planes of orders 3 and 4.
7. Prove Theorem 6. (This requires a few easy applications of Pasch's axiom.)
8. Prove Theorem 7. (Suppose $\mathbf{P}_3 = \mathbf{P}_3(P', \mathbf{P}_2')$. Prove $\mathbf{P}_3(P', \mathbf{P}_2') = \mathbf{P}_3(P', \mathbf{P}_2) = \mathbf{P}_3(P, \mathbf{P}_2)$, skipping special cases such as $P' \in \mathbf{P}_2$. Then do the special cases.)

9. Prove the analog of Theorem 3 for projective 3-space.
10. Prove that each subspace of a projective space is itself a projective space.

4. ANALYTIC PROJECTIVE SPACES

In the previous section we generalized the synthetic version of $\mathbf{P}_2\mathrm{R}$. Now we shall generalize the analytic (coordinatized) version. The dimension 2 will be replaced by an arbitrary integer $n \geq -1$. The real field R will be replaced by an arbitrary field F. The generalization is quite straightforward. The special importance of lines in the coordinatization of $\mathbf{P}_2\mathrm{R}$ will, of course, disappear for $n \neq 2$ (compare Section IV.1). Two m-tuples **a**, **b** of elements of a field F are said to be *proportional* if $\mathbf{a} = x\mathbf{b}$ for some x in F. Proportionality is an equivalence relation on the set F^{m*} of all m-tuples of elements of F except $\mathbf{0} = (0, 0, \cdots, 0)$ (Exercise 1).

DEFINITION 1 Let F be a field and n a nonnegative integer. *Projective n-space over F*, denoted by $\mathbf{P}_nF$, is the set of all equivalence classes, under proportionality, in F^{n+1*}. The elements of $\mathbf{P}_nF$, that is, these equivalence classes, are called *points*. $\mathbf{P}_{-1}F$ is defined to be the empty set.

Note that $\mathbf{P}_0F$ consists of just one point, namely, the class F^*. This is consistent with Definition 3.5(b).

In view of Theorem 2.1, we should be able to define lines, planes, and so on in $\mathbf{P}_nF$ by means of linear subspaces of F^{n+1}. The definitions, which are really only slight modifications of definitions in linear algebra, follow.

DEFINITIONS Let $P, P_1, P_2, \cdots, P_r$ be points in $\mathbf{P}_nF$, where F is a field, $n \geq -1$ an integer.

2. P is a *linear combination* of $P_1, \cdots, P_r$ if there are coordinate vectors $P{:}\mathbf{p}, P_1{:}\mathbf{p}_1, \cdots, P_r{:}\mathbf{p}_r$ such that $\mathbf{p}$ is a linear combination of $\mathbf{p}_1, \cdots, \mathbf{p}_r$.
3. $P_1, \cdots, P_r$ are *independent* (*dependent*) if $\mathbf{p}_1, \cdots, \mathbf{p}_r$ are linearly independent (respectively dependent).
4. The *subspace* $[M]$ *spanned* by a set M of (not necessarily independent) points is the set of all points in $\mathbf{P}_nF$ that are linear combinations of points in M.
5. A *basis* for a subspace **S** of $\mathbf{P}_nF$ is a set of linearly independent points that spans **S**.
6. If a subspace **S** of $\mathbf{P}_nF$ has a basis of r points, the *dimension* of **S**, written dim **S**, is $r - 1$. Subspaces of dimensions 1, 2, $n - 1$ are called, respectively, *lines*, *planes*, and *hyperplanes*.
7. Let **S** and **S**′ be subspaces of $\mathbf{P}_nF$. The *meet* $\mathbf{S} \wedge \mathbf{S}'$ of **S** and **S**′ is the intersection of the sets **S** and **S**′. The *join* $\mathbf{SS}'$ of **S** and **S**′ is the subspace $[\mathbf{S} \cup \mathbf{S}']$.

These definitions do not depend on choice of coordinate vectors. Suppose $P_i{:}\mathbf{p}_i$ and $P_i{:}\mathbf{p}'_i$, $1 \le i \le r$. Then $\mathbf{p}_i = b_i\mathbf{p}'_i$ for certain $b_i \neq 0$ in F. Any linear combination $\sum a_i\mathbf{p}_i$ of the $\mathbf{p}_i$ is then a linear combination $\sum a_ib_i\mathbf{p}'_i$ of the $\mathbf{p}'_i$. Also, of equal importance, a coefficient a_ib_i in the second representation is zero if and only if the corresponding coefficient a_i in the first representation is zero. With this remark, the following theorems are only slight modifications of theorems of linear algebra. The proofs are left as review exercises in linear algebra.

THEOREMS (Everything takes place in $\mathbf{P}_nF$, where F is a field, $n \ge -1$ an integer.)

1. If $Q_1, Q_2, \cdots, Q_s$ are points in $[P_1, P_2, \cdots, P_r]$, then $[Q_1, \cdots, Q_s]$ is a subspace of $[P_1, \cdots, P_r]$.
2. All bases of a given subspace have the same number of points. (Thus, Definition 6 is meaningful.)
3. $\dim \mathbf{P}_nF = n$.
4. If $\mathbf{S}$ and $\mathbf{S}'$ are subspaces, then $\mathbf{S} \wedge \mathbf{S}'$ and $\mathbf{SS}'$ are subspaces, and $\dim \mathbf{S} + \dim \mathbf{S}' = \dim \mathbf{SS}' + \dim \mathbf{S} \wedge \mathbf{S}'$.
5. If $\mathbf{S}_r$ is an r-dimensional subspace, then there exist $n + 1$ independent points $P_0, P_1, \cdots, P_r, Q_1, Q_2, \cdots, Q_{n-r}$ such that $P_i \in \mathbf{S}_r$, $0 \le i \le r$, and $Q_j \notin \mathbf{S}_r$, $1 \le j \le n - r$. These $n + 1$ points form a basis for $\mathbf{P}_nF$.
6. A subset $\mathbf{M}$ of $\mathbf{P}_nF$ is an r-dimensional subspace if and only if there exists an $(n + 1) \times (n + 1)$ matrix A over F of rank $n - r$ such that

$$\mathbf{M} = \{\text{points } P \mid P{:}\mathbf{p} \text{ and } \mathbf{p}A = \mathbf{0}\}.$$

As an example, we prove half of 6. Suppose that $\mathbf{M}$ is a subspace of $\mathbf{P}_nF$. Let $P_i{:}\mathbf{p}_i$, $0 \le i \le r$, be a basis of $\mathbf{M}$. Find $\mathbf{q}_j$'s such that $\{\mathbf{p}_0, \cdots, \mathbf{p}_r, \mathbf{q}_1, \cdots, \mathbf{q}_{n-r}\}$ is a basis of F^{n+1}. Define a linear transformation α of F^{n+1} by $\mathbf{p}_i\alpha = \mathbf{0}$, $0 \le i \le r$; $\mathbf{q}_j\alpha = \mathbf{q}_j$, $1 \le j \le n - r$. Let A be a matrix of α. Obviously, rank $A = n - r$. Now $P \in \mathbf{M}$ iff $P{:}\mathbf{p}$, where $\mathbf{p}$ is a linear combination of the $\mathbf{p}_i$. Say $\mathbf{p} = \sum a_i\mathbf{p}_i + \sum b_j\mathbf{q}_j$. Then $\mathbf{p}A = \sum b_j\mathbf{q}_j$, so $\mathbf{p}$ is a linear combination of the $\mathbf{p}_i$ iff $\mathbf{p}A = \mathbf{0}$. Thus, $P \in \mathbf{M}$ iff $\mathbf{p}A = \mathbf{0}$. ▮

In Chapter IV we shall deduce many other useful facts about $\mathbf{P}_nF$. The proofs of these facts will always be applications of linear algebra, although the terms used will not always be as close to those of linear algebra as in the above definitions.

Exercises

1. Let F be a field and m a positive integer. Prove that proportionality is an equivalence relation on F^{m*}. Is it an equivalence relation on F^m?

2. Let $\mathcal{P}$ be the set of all points and $\mathcal{L}$ the set of all lines in a $\mathbf{P}_nF$. Prove that $(\mathcal{P}, \mathcal{L}, \epsilon)$ is a projective space (Definition 3.1).
3. (See Exercise 2.) Prove that the planes in $(\mathcal{P}, \mathcal{L}, \epsilon)$ are the two-dimensional linear subspaces of $\mathbf{P}_nF$.
4. (See Exercise 2.) Prove that $(\mathcal{P}, \mathcal{L}, \epsilon)$ has dimension n, in the sense of Definition 3.5.
5. Prove that in $\mathbf{P}_3F$ every plane meets every line.
6. By Exercise 2 every $\mathbf{P}_2F$ can be regarded as a projective plane. Thus it has an *order* (Theorem 3.3). Find this order in terms of some number associated with F.
7. Prove Theorem 1.
8. Prove Theorem 2.
9. Prove Theorem 3.
10. Prove Theorem 4.
11. Prove Theorem 5. Note that $\{P_0, P_1, \cdots, P_r\}$ is a basis of $\mathbf{S}_r$.
12. Complete the proof of Theorem 6.
13. Generalize to n dimensions, and prove, the second sentence of Theorem 2.1.

5. MAPPINGS OF PROJECTIVE SPACES

We study mappings of projective spaces for two reasons. First, as in any theory of abstract systems, we need a way to tell when two geometries are "the same." Second, geometry is just the study of those properties of a space that are preserved by certain classes of mappings on the space. The basic property, which we must surely insist all of our mappings preserve, is that of incidence. Our most general definitions require only this.

DEFINITION 1 A *homomorphism* from a projective space $(\mathcal{P}, \mathcal{L}, \epsilon)$ to a projective space $(\mathcal{P}', \mathcal{L}', \epsilon')$ is a function μ from $\mathcal{P} \cup \mathcal{L}$ to $\mathcal{P}' \cup \mathcal{L}'$ such that: $\mathcal{P}\mu \subset \mathcal{P}'$; $\mathcal{L}\mu \subset \mathcal{L}'$; and, for all P in $\mathbf{P}$ and all l in $\mathcal{L}$, $P \epsilon l$ implies $P\mu \epsilon' l\mu$. A homomorphism that is one–one and onto is an *isomorphism*. An isomorphism from a space onto itself is a *collineation*.

The definition of isomorphism must be given an additional requirement in the case of one-dimensional spaces. This will be supplied in Section IV.4.

The correspondence between l and $\bar{l}$ mentioned in Theorem 3.1, together with the identity function on $\mathcal{P}$, provides an example of an isomorphism from an arbitrary projective space $\mathbf{S} = (\mathcal{P}, \mathcal{L}, \epsilon)$ onto the space $(\mathcal{P}', \mathcal{L}', \epsilon')$, where $\mathcal{P}' = \mathcal{P}$, $\mathcal{L}'$ is the set of all ranges of $\mathbf{S}$, and ϵ' is set membership. A less trivial example is the collineation $P_i\mu = P_{i+1 \,(\mathrm{mod}\, 7)}$, $l_i\mu = l_{i+1\,(\mathrm{mod}\, 7)}$ of the Fano plane shown in Figure II.3.1.

We begin by studying isomorphisms in general.

LEMMA 1 Let μ be an isomorphism from a projective space $\mathbf{S}$ to another projective space $\mathbf{S}'$. If A and B are distinct points of $\mathbf{S}$, then $(AB)\mu = A\mu B\mu$. If l and m are distinct intersecting lines of $\mathbf{S}$, then $(l \wedge m)\mu = l\mu \wedge m\mu$.

PROOF. $A \in AB$, $B \in AB$, and $A \neq B$ imply $A\mu \in' (AB)\mu$, $B\mu \in' (AB)\mu$, and $A\mu \neq B\mu$. Therefore, $(AB)\mu = A\mu B\mu$. $l \wedge m \in l$, $l \wedge m \in m$, and $l \neq m$ imply $(l \wedge m)\mu \in' l\mu$, $(l \wedge m)\mu \in' m\mu$, and $l\mu \neq m\mu$. Therefore, $(l \wedge m)\mu = l\mu \wedge m\mu$. ▮

LEMMA 2 Let μ be an isomorphism from a projective space $\mathbf{S}$ to another projective space $\mathbf{S}'$. Let P be a point and l a line of $\mathbf{S}$, with $P \notin l$. Then $P\mu \notin' l\mu$.

PROOF. Suppose on the contrary that $P\mu \in' l\mu$. Let $Q \in l$. Then $PQ \neq l$, but $(PQ)\mu = P\mu Q\mu = l\mu$, contradicting one–oneness of μ on lines. ▮

THEOREM 1 If a homomorphism μ of a projective plane $\mathbf{P} = (\mathcal{P}, \mathcal{L}, \epsilon)$ to a projective plane $\mathbf{P}' = (\mathcal{P}', \mathcal{L}', \epsilon')$ maps $\mathcal{P}$ one–one onto $\mathcal{P}'$, then it is an isomorphism of $\mathbf{P}$ onto $\mathbf{P}'$.

PROOF. We must show that μ maps $\mathcal{L}$ one–one onto $\mathcal{L}'$. Let l' be a line of $\mathcal{L}'$. Choose distinct points A', B' on l'. Since μ is onto $\mathcal{L}'$, there are points A, B (necessarily distinct) such that $A\mu = A'$, $B\mu = B'$. Then by Lemma 1 $(AB)\mu = l'$. Hence, μ is onto $\mathcal{L}'$. Now suppose μ is not one–one on $\mathcal{L}$. Then there are distinct lines l and m such that $l\mu = m\mu$. Let X be any point not on l. Let n be a line through X not through $l \wedge m$. Now $(n \wedge l)\mu \in l\mu$ and, since $m\mu = l\mu$, $(n \wedge m)\mu \in l\mu$. Then $n\mu = [(n \wedge l)(n \wedge m)]\mu = (n \wedge l)\mu(n \wedge m)\mu = l\mu$, so $X\mu \in l\mu$ for all X, contradicting the hypothesis that μ is onto $\mathcal{P}'$. ▮

THEOREM 2 An isomorphism maps subspaces onto subspaces of the same dimension.

PROOF. Let μ be an isomorphism of a space $\mathbf{S} = (\mathcal{P}, \mathcal{L}, \epsilon)$ onto a space $\mathbf{S}' = (\mathcal{P}', \mathcal{L}', \epsilon')$. Let $\mathbf{P}_r$ be an r-dimensional subspace of $\mathbf{S}$. We prove that $\mathbf{P}'_r = \mathbf{P}_r\mu$ is an r-dimensional subspace of $\mathbf{S}'$ by induction on r. This is trivial for $r = -1$ and 0. Suppose it is true for $r = k$. Let $\mathbf{P}_{k+1} = \mathbf{P}_{k+1}(P, \mathbf{P}_k)$ be a $(k + 1)$-dimensional subspace of $\mathbf{S}$. By the inductive hypothesis, $\mathbf{P}'_k = \mathbf{P}_k\mu$ is a k-dimensional subspace of $\mathbf{S}'$. P is not in $\mathbf{P}_k$; therefore, P is on no line in $\mathbf{P}_k$. Then by Lemma 2, $P' = P\mu$ is on no line in $\mathbf{P}'_k$, so P' is not in $\mathbf{P}'_k$. Therefore, $\mathbf{P}_{k+1}(P', \mathbf{P}'_k)$ is a $(k + 1)$-dimensional subspace of $\mathbf{S}'$. We complete the proof by showing that $\mathbf{P}'_{k+1} = \mathbf{P}_{k+1}\mu$ coincides with $\mathbf{P}_{k+1}(P', \mathbf{P}'_k)$. If Q is in $\mathbf{P}_{k+1}$ then $Q \in PS$ for some S in $\mathbf{P}_k$. By Lemma 1 then $Q\mu = Q' \in' P'S'$, where $S' = S\mu$ is in $\mathbf{P}'_k$, that is, Q' is in $\mathbf{P}_{k+1}(P', \mathbf{P}'_k)$. Thus, $\mathbf{P}'_{k+1} \subset \mathbf{P}_{k+1}(P', \mathbf{P}'_k)$. Conversely, let Q' be in $\mathbf{P}_{k+1}(P', \mathbf{P}'_k)$. Then $Q' \in' P'S'$ for some S' in $\mathbf{P}'_k$, and the preimage Q of Q' is on PS, where P, S are, respectively, the preimages of P', S'. Thus $Q \in \mathbf{P}_{k+1}$, and so $Q' \in \mathbf{P}'_{k+1}$. Hence, $\mathbf{P}'_{k+1} \supset \mathbf{P}_{k+1}(P', \mathbf{P}'_k)$. ▮

COROLLARY Isomorphisms preserve linear independence, linear dependence, and dimension.

As another nontrivial example of an isomorphism, we prove the following.

THEOREM 3 Every r-dimensional linear subspace of $\mathbf{P}_nF$ is isomorphic to $\mathbf{P}_rF$.

PROOF. Let $P_0, P_1, \cdots, P_r$ be a basis for the subspace $\mathbf{P}_r$. Let $\mathbf{p}_0, \cdots, \mathbf{p}_r$ be fixed coordinate vectors for the basis. For $P \in \mathbf{P}_r$ define

$$P\mu = \{(x_0, x_1, \cdots, x_r) \mid P\colon \sum x_i\mathbf{p}_i\}.$$

Now $P\colon \sum x_i\mathbf{p}_i$ if and only if $P\colon \sum (tx_i)\mathbf{p}_i$ for all $t \neq 0$. Hence, $P\mu \in \mathbf{P}_rF$ for every $P \in \mathbf{P}_r$. This also proves that μ is one–one (on points, and hence on lines). If $P'\colon(x_0, x_1, \cdots, x_r)$ is in $\mathbf{P}_rF$, let P be the point in $\mathbf{P}_nF$ with coordinate vector $\sum x_i\mathbf{p}_i$. Then $P \in \mathbf{P}_r$ and $P\mu = P'$. Hence, μ is onto. By our previous results we need only show that μ preserves incidence. However, to illustrate the situation further, we show directly that μ preserves independence of points. To this end, let $Q_1, Q_2, \cdots, Q_k$ be points in $\mathbf{P}_r$ with respective coordinate vectors $\mathbf{q}_1, \cdots, \mathbf{q}_k$. Let $\mathbf{q}'_1, \cdots, \mathbf{q}'_k$ be, respectively, coordinate vectors for $Q_1\mu, \cdots, Q_k\mu$. Say $\mathbf{q}'_i = (q_{i0}, q_{i1}, \cdots, q_{ir})$, $1 \leq i \leq k$. Then, by the definition of μ, we have

$$\mathbf{q}_i = \sum_j q_{ij}\mathbf{p}_j, \qquad 1 \leq i \leq k.$$

We must show that $\mathbf{q}_1, \cdots, \mathbf{q}_k$ are independent if and only if $\mathbf{q}'_1, \cdots, \mathbf{q}'_k$ are independent. Suppose the $\mathbf{q}_i$ are independent. Suppose for some a_i that $\sum a_i\mathbf{q}'_i = \mathbf{0}$. Then, $\sum a_iq_{ij} = 0$, $0 \leq j \leq r$, so that

$$\sum_i a_i\mathbf{q}_i = \sum_i a_i \sum_j q_{ij}\mathbf{p}_j = \sum_j \Big(\sum_i a_iq_{ij}\Big)\mathbf{p}_j = \mathbf{0}.$$

Then, by independence of the $\mathbf{q}_i$, $a_i = 0$ for all i. Hence, the $\mathbf{q}'_i$ are independent. Conversely, suppose the $\mathbf{q}'_i$ are independent. Suppose for some a_i that $\sum a_i\mathbf{q}_i = \mathbf{0}$. Then

$$\mathbf{0} = \sum_i a_i \sum_j q_{ij}\mathbf{p}_j = \sum_j \Big(\sum_i a_iq_{ij}\Big)\mathbf{p}_j,$$

and, by independence of the $\mathbf{p}_j$, $\sum a_iq_{ij} = 0$ for all j. Hence, $\sum a_i\mathbf{q}'_i = \mathbf{0}$, and, by independence of the $\mathbf{q}'_i$, $a_i = 0$ for all i. Therefore, the $\mathbf{q}_i$ are independent. ∎

We turn now to the most important mappings of the next chapter, namely, collineations of a projective plane.

The most familiar collineations of the real projective plane are, of course, those that come from the basic geometrical transformations of $\mathbf{E}_2$:

1. Translations.
2. Stretchings or shrinkings (collectively called homologies).
3. Reflections.

Items (1) and (2) extend to $\mathbf{P}_2\mathrm{R}$ in a very nice way [Figures II.5.1 (a) and (b)]; they fix every point on the new line. Note that these collineations

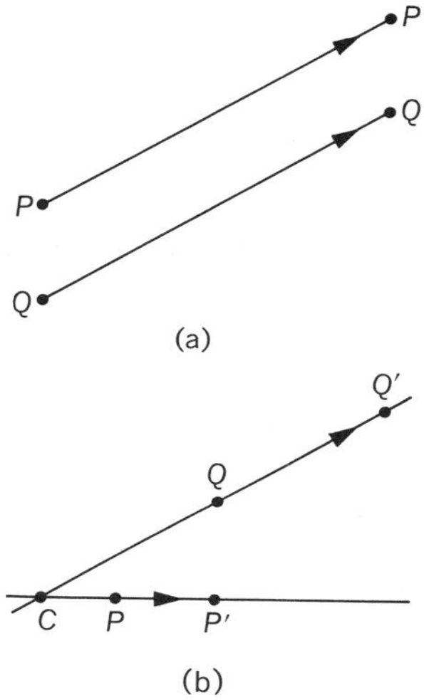

Fig. II.5.1

also fix (dually) every line through a certain point. In Figure II.5.1(b) the point is C, the center of the homology; in (a) it is the new point common to lines QQ', PP', and so on. This prompts the following.

DEFINITION 2 Let π be a collineation of a projective plane $\mathbf{P}$. A point C of $\mathbf{P}$ is a *center* of π if $l\pi = l$ for every line l through C. A line a of $\mathbf{P}$ is an *axis* of π if $P\pi = P$ for every point P on a.

Note that a collineation fixes its center (or axis), if it has one. Also, beware of the following: A collineation may fix a line but not fix the points on the line (for example, PP' in Figure II.5.1). The next two theorems show that collineations of this type have the same basic properties in any projective plane that they have in the real projective plane.

THEOREM 4 If a collineation of a projective plane has a center, then it has an axis, and conversely.

PROOF. Suppose the collineation π has center C.

CASE 1. $a\pi = a$ for some line a not through C. Then, for each $P \in a$, $P\pi = (CP \wedge a)\pi = CP \wedge a = P$. Hence, a is an axis of π.

CASE 2. $l\pi \neq l$ for every line l not through C. Choose l_1 not containing C. Let $L_1 = l_1 \wedge l_1\pi$. Then $L_1\pi = (CL_1 \wedge l_1)\pi = CL_1 \wedge l_1\pi = L_1$. Let $a = CL_1$. Suppose that $P \in a$, $P \neq C$, and $P \neq L_1$. Choose l_2 such that $P \in l_2$ and $C \notin l_2$. As with L_1, point $L_2 = l_2 \wedge l_2\pi$ is fixed by π. If L_2 were not on a, L_1L_2 would be a line not through C which is fixed by π. Therefore, $L_2 \in a$, and so $P = L_2$ is fixed by π. Hence, a is an axis of π. The converse statement is dual. ∎

THEOREM 5 A nonidentity collineation of a projective plane has at most one center and one axis.

PROOF. Suppose a collineation π of a projective plane **P** has distinct centers C_1 and C_2. If P is a point not on C_1C_2, then $P\pi = (PC_1 \wedge PC_2)\pi = PC_1 \wedge PC_2 = P$. If Q is a point on C_1C_2, let l be a line through Q distinct from C_1C_2 and let Q_1 and Q_2 be points of l, with Q, Q_1, Q_2 distinct. (Every line has at least three points!) Then $Q\pi = (Q_1Q_2 \wedge C_1C_2)\pi = Q_1Q_2 \wedge C_1C_2 = Q$. Thus, π fixes every point, and, hence, every line, of **P**. The statement for axes is dual. ∎

A collineation with center C and axis a is called a *(C, a)-perspectivity*. The discussion preceding Definition 2 gives rise to Definition 3.

DEFINITION 3 A (C, a)-perspectivity π of a projective plane is an

*elation,** if π is the identity or if C is on a;
homology, if π is the identity or if C is not on a.

Collineations of Type (3), as well as those arising from 180° rotations, motivate another definition.

DEFINITION 4 A collineation μ is *involutory* (and is called an *involution*) if μ is not the identity, but μ^2 is the identity.

We shall need this definition in Chapter VI.

Exercises

1. Prove that the set of all collineations of a projective space forms a group under composition of functions.
2. Let π be a (C, a)-perspectivity and μ an arbitrary collineation of a projective plane. Prove that $\mu^{-1}\pi\mu$ is a $(C\mu, a\mu)$-perspectivity of the projective plane.
3. Let π be a perspectivity of a projective plane of finite order n. Suppose π,

* Some authors say *translation*.

considered as an element of the group of collineations (see Exercise 1), is of order s. Prove that π is an elation if and only if s divides n, and a homology if and only if s divides $n - 1$.

*4. Find the group of all collineations of the Fano plane (Figure II.3.1).
5. Are all collineations of the Fano plane perspectivities?
6. Prove that any two planes in a projective 3-space are isomorphic. (Hint: Use projection.)
7. Let ν be an involutory (C, a)-elation of a projective plane of finite even-order $n = 2m$. Show that ν permutes the n^2 lines not through C in $n^2/2 = 2m^2$ classes of two lines each. Show that ν permutes the n points different from C on a line through C different from a in $n^2/2 = 2m^2$ classes of two points each.

*8. Extend Theorem 1 to finite dimensional spaces; to general spaces.
9. Find collineations of $\mathbf{P}_2\mathrm{R}$ that are not perspectivities.

6. AFFINE SPACES AND OTHER TOPICS

If we use the basic incidence properties of $\mathbf{E}_2$ to create an abstract geometrical structure, we are led to the following definition.

DEFINITION 1 An *affine plane* is a triple $(\mathcal{P}, \mathcal{L}, \epsilon)$ of sets, where $\mathcal{P} \wedge \mathcal{L} = \emptyset$ and $\epsilon \subset \mathcal{P} \times \mathcal{L}$ (interpreted as in Section 3), satisfying:

(1) Two distinct points are on exactly one line.
(2) If l is a line and P a point not on l, then there is exactly one line through P that does not intersect l.
(3) There are four points, no three collinear.

The connection between affine and projective planes in general is the same as the connection made between $\mathbf{E}_2$ and $\mathbf{P}_2\mathrm{R}$ in Section 1.

THEOREM 1

(a) If one line and all the points on it are removed from a projective plane, the remaining incidence structure is an affine plane.

(b) Given an affine plane, there is a projective plane that determines it as in (a).

PROOF

(a) If line l_∞ is removed, then $P(l \wedge l_\infty)$ is the unique line required by Condition 2. Conditions 1 and 3 are clear.

(b) Let $\mathbf{A} = (\mathcal{P}, \mathcal{L}, \epsilon)$ be an affine plane. Call two lines l, m of $\mathbf{A}$ *parallel* if either $l = m$ or l and m have no common points. This defines an equivalence relation on $\mathcal{L}$. Add one new point to $\mathcal{P}$ for each equivalence class of parallel lines to form the set $\mathcal{P}'$. Add one new line l_∞ to $\mathcal{L}$ to form the set $\mathcal{L}'$. Define $\epsilon' \subset \mathcal{P}' \times \mathcal{L}'$ as follows: If $P \in \mathcal{P}$ and $l \in \mathcal{L}$, then $P \epsilon' l$ iff $P \epsilon l$; if P is new and $l \in \mathcal{L}$, then $P \epsilon' l$ iff P corresponds to the equivalence class containing l; $P \epsilon' l_\infty$ iff P is new. It is easy to check that $(\mathcal{P}', \mathcal{L}', \epsilon')$ is a projective plane and that if l_∞ and all its points are removed, then **A** remains. ▎

COROLLARY Associated with every affine plane **A** is a cardinal number n, called the *order* of **A**, such that the following holds.

(1) Each line of **A** contains exactly n points.
(2) Each point of **A** is on exactly $n + 1$ lines.
(3) **A** contains exactly n^2 points.
(4) **A** contains exactly $n^2 + n$ lines.

PROOF. Proof is seen immediately from (b) and Theorem 3.4. ▎

If we generalize $\mathbf{E}_2$ as a *coordinatized* plane, we are led to a second definition.

DEFINITION 2 Let F be a field. The points of the *analytic affine plane* $\mathbf{A}_2F$ are the ordered pairs of elements of F. Lines are point sets of two types.

Nonvertical: $\{(x, y) \in F \times F \mid y = mx + b\}$, where m and b are fixed elements of F.

Vertical: $\{(x, y) \in F \times F \mid x = a\}$, where a is a fixed element of F.

THEOREM 2 Let F be a field. Any affine plane obtained from $\mathbf{P}_2F$ by the removal of a line and all its points is isomorphic* to $\mathbf{A}_2F$.

PARTIAL PROOF. Suppose **A** is obtained from $\mathbf{A}_2F$ by removal of l_∞. l_∞ is arbitrary, but it follows from Section IV.4 that we can assume $l_\infty\colon [0, 0, 1]$. Then a point P is on l_∞ if and only if $P\colon (a, b, 0)$ for some a, b. Hence, every point of **A**, that is, every point not on l_∞, has a unique coordinate of the form $(x, y, 1)$. The isomorphism on points is defined by $(x, y, 1) \rightarrow (x, y)$. Every line other than l_∞ either has a unique coordinate of the form $[m, -1, b]$ or a unique coordinate of the form $[-1, 0, a]$. The isomorphism on lines is defined by $[m, -1, b] \rightarrow \{(x, y) \mid y = mx + b\}$, $[-1, 0, a] \rightarrow \{(x, y) \mid x = a\}$. Preservation of incidence is easily checked. ▎

It follows from Theorem 2 that any two affine planes obtained from $\mathbf{P}_2F$ are isomorphic. We shall see later that this is not always the case for general projective planes.

We relegate the higher-dimensional version of these results to the exercises. Definitions and results concerned with other abstract geometrical systems also appear there. The following exercises constitute an important part of this section; the student should at least read all of them.

* Definition 5.1 clearly makes sense for affine planes.

Exercises

1. Show that (3) of Definition 1 can be replaced by the weaker requirement that there be three noncollinear points.
2. Show that the projective plane of Theorem 1(b) is uniquely determined, up to isomorphism, by the given affine plane.
3. Show that an analytic affine plane (Definition 2) is an affine plane (Definition 1).
4. The Veblen–Young axiom, the key to our definition of projective space, was derived from the statement that "any two lines in the same plane intersect," which we felt should be true in a projective space. The corresponding statement for affine space is, "Given a point P not on a line l, all but exactly one of the lines that are in the plane of P and l and that pass through P intersect l." Proceeding from this, give a definition of *affine space*. Then, following Section 3, define *affine subspace* and *dimension*. Finally, state and prove a version of Theorem 1 for affine spaces of arbitrary finite dimension.
5. For F a field and $n \geq -1$ an integer define *analytic affine n-space* $\mathbf{A}_n F$. Following Section 4, define subspace, dimension, and so forth. Check that $\mathbf{A}_n F$ is an affine space in the sense of Exercise 3.
6. State and prove (partially) the n-dimensional version of Theorem 2. (Here you may assume that the hyperplane to be removed is $\{(x_0, x_1, \cdots, x_n) \mid x_0 = 0\}$.)

* * *

All geometries studied in this book are examples of the following.

DEFINITION 3 A *partial plane** is a triple $(\mathcal{P}, \mathcal{L}, \epsilon)$, interpreted as usual, such that two distinct points are on at most one line.

7. Show that in a partial plane two distinct lines intersect in at most one point.
8. Is every subset of a projective space a partial plane?
9. A partial plane $\Pi = (\mathcal{P}, \mathcal{L}, \epsilon)$ is a *subplane* of a partial plane $\Pi' = (\mathcal{P}', \mathcal{L}', \epsilon')$, and we write $\Pi \subset \Pi'$, if $\mathcal{P} \subset \mathcal{P}'$, $\mathcal{L} \subset \mathcal{L}'$, and $\epsilon \subset \epsilon'$. [Note that we do not require $\epsilon = \epsilon' \cap (\mathcal{P} \times \mathcal{L})$. Give an example where this equation does not hold.] The class of partial planes is partially ordered by this relation. Discuss the possibility of two partial planes having a greatest lower bound or least upper bound relative to this order.

* * *

---

* There is little justification for the term *plane* here.

Affine planes have the same basic incidence structure as $\mathbf{E}_2$; they could be called "generalized Euclidean planes." Projective planes are a type of "generalized non-Euclidean plane"; they are "non-Euclidean" in the classical sense that Postulate (2) of Definition 1 is denied. But there is another way to deny this postulate. We can insist that for every point P and line l with P not on l there are two or more lines through P that do not intersect l. This postulate, called the *Bólyai–Lobachevsky* postulate for the men who first studied it successfully, is more difficult to use than the corresponding projective and affine postulates. For example, this postulate is obviously satisfied in $\mathbf{E}_3$, but we surely do not want to think of $\mathbf{E}_3$ as a Bólyai–Lobachevsky plane. We need a fairly complicated "limiting postulate" to get a sensible structure. Here is one way to do this.

DEFINITION 4 (L. M. Graves.) A *Bólyai–Lobachevsky plane** is a partial plane $\mathbf{B}$ such that:

(1) Any two points are on a line.
(2) The Bólyai–Lobachevsky postulate holds.
(3) Each line has the same number of points.
(4) There are four points, no three collinear.
(5) (Limiting postulate.) If $(\mathcal{P}, \mathcal{L}, \epsilon)$ is a subplane of $\mathbf{B}$ such that (a) $P \neq Q$ in $\mathcal{P} \rightarrow PQ \in \mathcal{L}$, and (b) $P \epsilon l \in \mathcal{L} \rightarrow P \in \mathcal{P}$, then $(\mathcal{P}, \mathcal{L}, \epsilon) = \mathbf{B}$.

*10. Find a finite Bólyai–Lobachevsky plane. (There is one with 13 points and 26 lines.)

11. Suppose that in a finite Bólyai–Lobachevsky plane each line is on exactly π points and each point is on exactly λ lines. Prove that the plane has exactly $1 + \lambda(\pi - 1)$ points and exactly $\frac{\lambda}{\pi}[1 + \lambda(\pi - 1)]$ lines.

* * *

Another type of geometric structure, not a partial plane, is based on the fact that three distinct noncollinear points in $\mathbf{E}_2$ determine a unique circle.

DEFINITION 5 A *Moebius plane* is a triple of sets $(\mathcal{P}, \mathcal{C}, \epsilon)$ with $\mathcal{P} \cap \mathcal{C} = \emptyset$, $\epsilon \subset \mathcal{P} \times \mathcal{C}$ ($\mathcal{P}$: "points," $\mathcal{C}$: "circles," ϵ: "on") such that:

(1) Any three distinct points are on exactly one circle.
(2) Given any circle c and points $P \epsilon c$ and $Q \not\epsilon c$, there is a unique circle d such that $Q \epsilon d$ and P is the only point on both c and d.
(3) There are four points, not all on the same circle (that is, not concyclic).

---

* Also called a *hyperbolic plane*.

12. Moebius' name appears in Definition 5 because, as you must show, the following is an example of such a plane. $\mathcal{P} = \mathrm{C} \cup \{\infty\}$, the field of complex numbers with a symbol "∞" adjoined. The "circle" determined by three distinct points P, Q, R is the set of all points X such that $(PQRX) \in \mathrm{R} \cup \{\infty\}$, where R is the field of real numbers. The symbol $(PQRX)$ is defined for distinct complex numbers to be the number $(P - Q)(R - X)(P - X)^{-1}(R - Q)^{-1}$. The use of the symbol ∞ is easy to understand. For example, $(PQRP) = \infty$, $(PQR\infty) = (P - Q)(R - Q)^{-1}$. (Hint: Look up "Moebius transformation" or "cross ratio" in books on complex analysis.)
13. Let P be a fixed point in the Moebius plane $\mathbf{M} = (\mathcal{P}, \mathcal{C}, \epsilon)$. Prove that $\mathbf{M}_P = (\mathcal{P} - \{P\}, \{c \in \mathcal{C} \mid P \epsilon c\}, \epsilon$ restricted) is an affine plane.
14. Suppose some circle of a Moebius plane is on exactly $n + 1$ points. (If you prefer, assume that n is finite.) Prove:

    (a) Every circle is on exactly $n + 1$ points.
    (b) Every point is on exactly $n^2 + n$ circles.
    (c) There are exactly $n^2 + 1$ points.
    (d) There are exactly $n^3 + n$ circles.
    (e) Two distinct points are on exactly (how many?) common circles.

DEFINITION 6 An *ovoid* in a projective n-space is a subset $\mathbf{O}$ of the space such that:

(1) No three points of $\mathbf{O}$ are collinear.
(2) If P is any point on $\mathbf{O}$, then the set of all points that are on lines intersecting $\mathbf{O}$ only at P is precisely the set of all points of a hyperplane.

15. Let $\mathbf{O}$ be an ovoid in a projective 3-space. Let $\mathbf{M(O)} = (\mathbf{O}, \{\mathbf{P} \wedge \mathbf{O} \mid \mathbf{P}$ is a plane containing more than one point of $\mathbf{O}\}, \in)$. Prove that $\mathbf{M(O)}$ is a Moebius plane.

* * *

Still another class of generalizations of projective planes is the class of *combinatorial designs*. An important subclass of this, which finds application in the field of statistics known as "design of experiments," is the class of *balanced incomplete block designs*. One of the most important subclasses of this class is defined as follows:

DEFINITION 7 A (v, k, λ)-*configuration* is a family of v sets $S_1, S_2, \cdots, S_v$ such that:

(1) Each S_i is a k-element set.
(2) $S = S_1 \cup S_2 \cup \cdots \cup S_v$ is a v-element set.

(3) For each $i \neq j$, $S_i \cap S_j$ is a λ-element set.
(4) (nontriviality) $0 < \lambda < k < v - 1$.

16. For what values of the parameters v, k, λ is $(S, \{S_i\}, \in)$ a projective plane? An affine plane?
17. Prove that in any (v, k, λ)-configuration, $k(k - 1) = \lambda(v - 1)$.
*18. Prove that in a (v, k, λ)-configuration where v is even, $k - \lambda$ must be a perfect square. [Hint: Let $S = \{p_1, p_2, \cdots, p_v\}$. Let $a_{ij} = 1$ if $x_i \in S_j$, $a_{ij} = 0$ otherwise. Let A be the matrix (a_{ij}). Calculate the determinant of AA^T, where A^T is the transpose of A.]

Another interesting type of balanced incomplete block design is the following.

DEFINITION 8 A *Steiner triple system* of order v is a family of sets $S_1, S_2, \cdots, S_b$ such that:

(1) Each S_i is a 3-element set.
(2) $S = S_1 \cup S_2 \cup \cdots \cup S_b$ is a v-element set.
(3) Each 2-element subset of S is contained in exactly one S_i.

19. The Fano plane provides, in an obvious way, an example of a Steiner triple system of order 7.
*20. The only permissible orders $v \leq 10$ of Steiner triple systems are $v = 3, 7, 9$. Systems of these orders exist and are unique.

* * *

A student interested in an extremely general approach to geometric structures could consult the first few pages of [18], although to understand the main content of this reference requires considerable background in algebra.

III

Synthetic Versus Analytic Projective Spaces

1. COORDINATIZATION OF AN ARBITRARY PROJECTIVE PLANE

In this section we shall introduce coordinates for the points and lines of an arbitrary projective plane (Definition II.3.3). Naturally we will be guided by our work in the more familiar spaces (Section II.1) but, due to the rather great generality permitted by the axioms of projective planes, the task is not a simple one. Perhaps the most troublesome thing is that we cannot begin by assigning elements of some algebraic system, such as a field, to points and lines; but rather we must create the underlying system from the coordinates assigned. Furthermore, the algebraic system finally constructed is necessarily, it appears, unlike the systems commonly discussed in algebra. However, the results finally obtained are well worth the effort!

There are several other schemes for coordinatization of a projective plane, each differing slightly from the one below. (See Exercise VI.6.6 for an example.) When consulting other books or papers, the reader may often have to begin by learning what coordinatization is being used.

*   *   *

Let **P** be a projective plane. Let g, h, and l_∞ be three nonconcurrent lines in **P**. We shall try to coordinatize the points of **P** not on l_∞ rather like the points of the real projective plane not on the new line. Say we want g to become the x axis; h the y axis. Then $O = g \wedge h$ should be the origin.

Let S be a set of cardinality n, where n is the order of **P** (Theorem II.3.3). We assume S contains the symbols 0 and 1 but not the symbol ∞. Let $X = g \wedge l_\infty$. Assign the ordered pairs $(x, 0)$, $x \in S$, in an arbitrary one–one way to the points other than X on g, except be sure to assign $(0, 0)$ to O. In $\mathbf{E}_2$ we could now coordinatize the points on the y axis by means of a

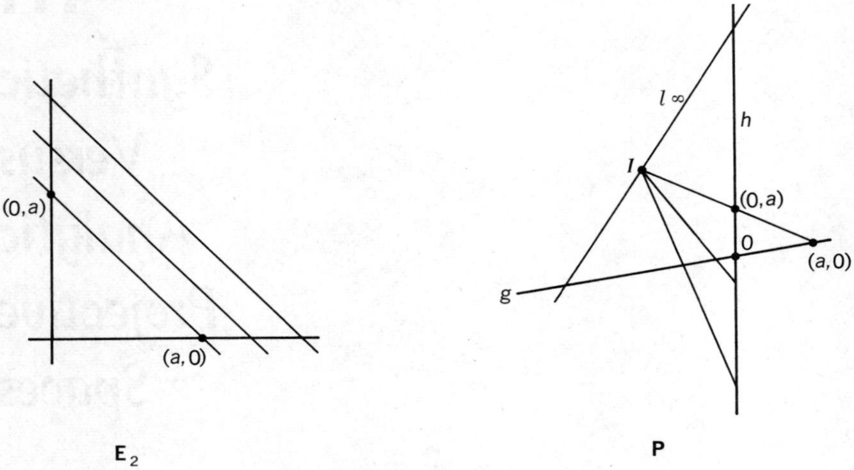

Fig. III.1.1

family of parallel lines (Figure III.1.1). In **P** we do the corresponding thing—select a point I different from X and Y on l_∞. If P is on h, $P \neq Y$, then $PI \wedge g$ has been assigned coordinates, say $(x, 0)$. Assign P the coordinates $(0, x)$. Note that this reassigns $O\!:\!(0, 0)$.

Coordinatization of all other points in $\mathbf{E}_2$ could now be effected in the usual way (Figure III.1.2). In **P**: If P is a point not on l_∞ say $PY \wedge g\!:\!(a, 0)$, $PX \wedge h\!:\!(0, b)$. Then let $P\!:\!(a, b)$. Note that this procedure assigns the coordinates already given to points on g and h.

All lines through a fixed point on l_∞ should have the same slope. This slope can be used as a coordinate for the point. In $\mathbf{E}_2$ the line through $(1, 0)$ and

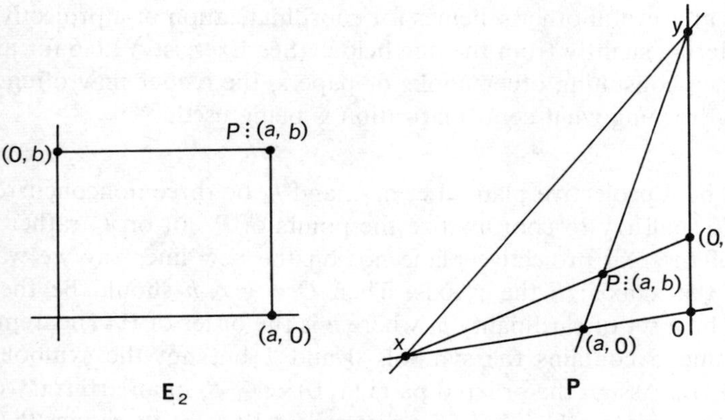

Fig. III.1.2

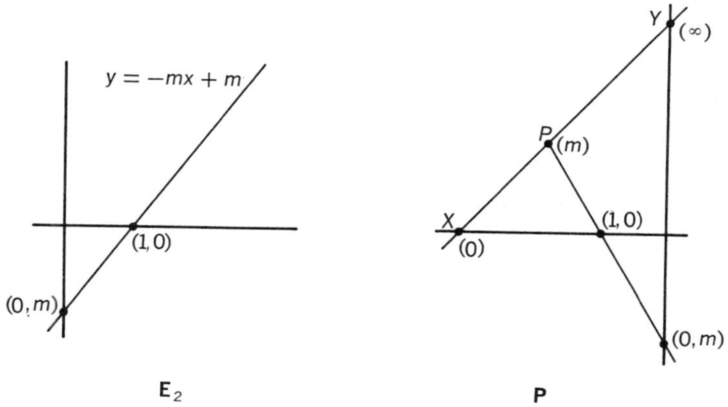

Fig. III.1.3

(0, m) has slope $-m$. In **P** we have as yet no algebraic structure, so we will neglect the minus sign and proceed as shown in Figure III.1.3. This assigns a coordinate $P\colon(m)$, $m \in S$, in a one–one manner to every point on l_∞ except Y. Note that $X\colon(0)$ and $I\colon(1)$. Finally, let $Y\colon(\infty)$.

Now to coordinatize the lines of **P**. We do nothing to l_∞. Any line k other than l_∞ that passes through Y will intersect $g = OX$ in a point having coordinates of the form $(a, 0)$. Let $k\colon[a]$. Any line l not through Y will intersect l_∞ in a point (m) and $h = OY$ in a point $(0, b)$. Then, let $l\colon[m, b]$.

In $\mathbf{E}_2$, a line with slope $-m$ that passes through (a, b) has y-intercept $mx + y$. In **P**, we define a function T from the set S^3 of all ordered triples of elements of S into S as follows (Figure III.1.4). $\mathsf{T}(m, a, b) = c$ if and only

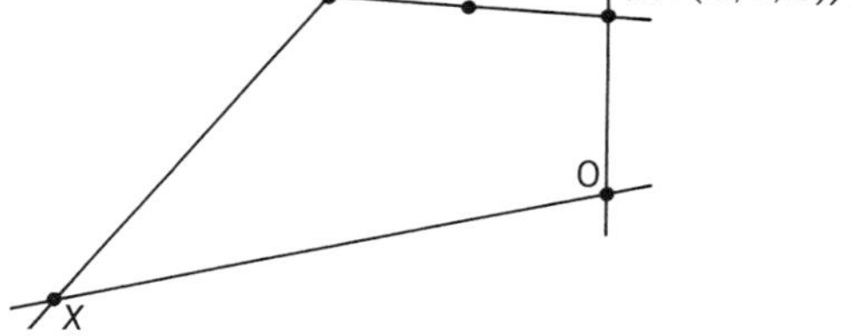

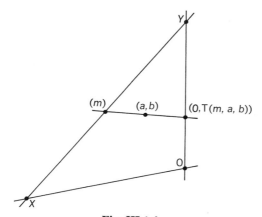

Fig. III.1.4

if (m), (a, b), and $(0, c)$ are collinear. It is clear from Figure III.1.4 that T is a function with domain S^3 and range S.

THEOREM 1 In a projective plane, coordinatized as above, a point (x, y) is on a line $[m, a]$ if and only if $\mathsf{T}(m, x, y) = a$.

PROOF. If (x, y) is on $[m, a]$, then $(0, a) = (m)(x, y) \wedge [0] = (0, \mathsf{T}(m, x, y))$, so $\mathsf{T}(m, x, y) = a$. Conversely, if $\mathsf{T}(m, x, y) = c$, then $(m)(x, y) \wedge [0] = (0, a)$, and so $(m)(x, y) \vdots [m, a]$. Hence, (x, y) is on $[m, a]$. ∎

$\mathsf{T}(a, b, c)$ is then an analog of the expression $ab + c$ in a field. We shall use T to give S its algebraic structure. To this end we now find some properties enjoyed by T in every projective plane.

THEOREM 2 Let a projective plane be coordinatized by a set S and let T be the ternary operation on S, as defined above. Then:

(1) $\mathsf{T}(a, 0, c) = \mathsf{T}(0, b, c) = c$ for all a, b, c in S.
(2) $\mathsf{T}(a, 1, 0) = \mathsf{T}(1, a, 0) = a$ for all a in S.
(3) For any (a, b, c) in S^3 there exists exactly one x in S such that $\mathsf{T}(a, b, x) = c$.
(4) For any (a, b, c, d) in S^4 with $a \neq c$ there exists exactly one x in S such that

$$\mathsf{T}(x, a, b) = \mathsf{T}(x, c, d).$$

(5) For any (a, b, c, d) in S^4 with $a \neq c$ there exists exactly one (x, y) in S^2 such that

$$\mathsf{T}(a, x, y) = b \quad \text{and} \quad \mathsf{T}(c, x, y) = d.$$

PROOF

(1) $(a)(0, c) \wedge [0] = (0)(b, c) \wedge [0] = (0, c)$.
(2) From the way points on l_∞ were coordinatized, we have $(a)(1, 0) \wedge [0] = (0, a)$. From the way points on $[0]$ were coordinatized, we have $(1)(a, 0) \wedge [0] = (0, a)$.
(3) x must be such that (b, x) is on $[a, c]$. But (b, x) is also on $[b]$, and $[b]$ intersects $[a, c]$ at exactly one point. Thus, x exists and is unique.
(4) x must be such that $(x)(a, b) \wedge [0] = (x)(c, d) \wedge [0]$. Thus, $(x) = (a, b)(c, d) \wedge l_\infty$, proving uniqueness. x is in S, because $x = \infty$ would imply $a = c$.
(5) (x, y) must be on $[a, b]$ and $[c, d]$. Since $a \neq c$, $[a, b] \wedge [c, d]$ is unique and not on l_∞. Thus, (x, y) exists and is unique. ∎

DEFINITION 1 (M. Hall.) A *planar ternary ring* is a pair (S, T) where S is a set containing 0 and 1, but not ∞, and T is a function from S^3 to S satisfying (1) through (5) of the preceding theorem.

Theorem 2 states that every coordinatization of every projective plane produces a planar ternary ring. In the converse direction we have Theorem 3.

THEOREM 3 Given a planar ternary ring, there is a projective plane **P** and a coordinatization of **P** whose resulting planar ternary ring is the one given.

PARTIAL PROOF. We will show how to define the plane $\mathbf{P} = (\mathcal{P}, \mathcal{L}, \epsilon)$ given the ring (S, T). It will remain for the student to show that **P** is a projective plane and to obtain (S, T) from **P** (Exercise 1). Let $\mathcal{P} = S^2 \cup \{(x) \mid x \in S\} \cup \{(\infty)\}$. Let $\mathcal{L} = \{[a, c] \mid (a, c) \in S^2\} \cup \{[a] \mid a \in S\} \cup \{l_\infty\}$. Define $\epsilon \subset \mathcal{P} \times \mathcal{L}$ as follows:

$$\begin{aligned}
&(x, y) \in [a, c] \text{ iff } \mathsf{T}(a, x, y) = c,\\
&(x, y) \in [a] \text{ iff } x = a,\\
&(x, y) \notin l_\infty \text{ for any } x \text{ and } y,\\
&\quad (x) \in [a, c] \text{ iff } x = a,\\
&\quad (x) \in [a] \text{ iff } x = \infty,\\
&\quad (x) \in l_\infty \text{ for all } x.
\end{aligned}$$ ❙

DEFINITION 2 In a planar ternary ring (S, T) *addition* and *multiplication* are binary operations defined for all x and y in S as follows:

$$x + y = \mathsf{T}(1, x, y) \qquad xy = \mathsf{T}(x, y, 0).$$

As we shall see in Section 3, it is not necessarily true that $\mathsf{T}(x, y, z) = xy + z$. This is the reason that we must work with the ternary operation at first.

THEOREM 4 If (S, T) is a planar ternary ring, then $(S, +)$ and $(S - \{0\}, \cdot)$ are loops with identities 0 and 1, respectively. (See Section I.3 for definitions.)

PROOF. S is obviously closed under $+$, $x + 0 = \mathsf{T}(1, x, 0) = x$ by (2); $0 + x = \mathsf{T}(1, 0, x) = x$ by (1). For each (a, b) in S^2 there is, by (3), a unique x such that $a + x = \mathsf{T}(1, a, x) = b$. By (5) there is a unique (y, u) in S^2 such that $\mathsf{T}(1, y, u) = b$ and $u = \mathsf{T}(0, y, u) = a$, that is, there is a unique y such that $y + a = \mathsf{T}(1, y, a) = b$. This proves that $(S, +)$ is a loop with identity 0.
Let $a \in S - \{0\}$, $b \in S$. By (5) there is a unique (x, y) in S^2 such that $\mathsf{T}(a, x, y) = b$ and $y = \mathsf{T}(0, x, y) = 0$, that is, there is a unique x such that $ax = \mathsf{T}(a, x, 0) = b$. By (1), $a \cdot 0 = 0$. Then the unique solution to $ax = 0$ with $a \neq 0$ is $x = 0$. Hence, if $ax \neq 0$ and $a \neq 0$, then $x \neq 0$. Now, by (4) there exists a unique y such that $ya = \mathsf{T}(y, a, 0) = \mathsf{T}(y, 0, b) = b$. By (1) $0 \cdot a = 0$. Hence, in the equation $ya = b$, $y = 0$ if $b = 0$ and $y \neq 0$ if $b \neq 0$. We have shown that if a and b are in $S - \{0\}$ then $ab \in S - \{0\}$ and $ax = b$, $ya = b$ have unique solutions x, and y in $S - \{0\}$. Finally, by (2), $1 \cdot a = a \cdot 1 = a$ for all a. ❙

How does the coordinatization procedure described above appear in $\mathbf{P}_2F$, where F is a field? The reader may check (Exercise 2) that if we choose for O, X, Y, I the points with coordinates $(0, 0, 1)$, $(1, 0, 0)$, $(0, 1, 0)$, $(1, 1, 0)$, respectively, and if we assign to the point $(x, 0, 1)$, different from X on line OX, the coordinate $(x, 0)$, then the following points and lines will be assigned the following coordinates:

$$\begin{aligned}(x, y, 1)&:(x, y),\\ (-1, x, 0)&:(x),\\ (0, 1, 0)&:(\infty),\\ [a, 1, b]&:[a, b],\\ [-1, 0, a]&:[a],\\ [0, 0, 1]&:l_\infty.\end{aligned}$$

This should be compared with the discussion of affine plane coordinates in Section II.6.

Exercises

1. Complete the proof of Theorem 3.
2. Verify the statements made after the proof of Theorem 4.
3. Coordinatize the Fano plane (Figure II.3.1). Then exhibit the Fano plane as a $\mathbf{P}_2F$, where F is the field of two elements. Illustrate the remarks made after the proof of Theorem 4 by this case.
4. Same as Exercise 3 for the plane of order three and the field of three elements.
5. Let $\mathbf{P}$ be a projective plane of finite order n. Let $N = n^2 + n + 1$. Let $P_1, P_2, \cdots, P_N$ be the points; $l_1, l_2, \cdots, l_N$ the lines of P. The *incidence matrix* $A = (a_{ij})$ is the $N \times N$ matrix defined by $a_{ij} = 1$ if $P_i \in l_j$, $a_{ij} = 0$ if $P_i \notin l_j$. Describe the matrices AA^T and A^TA. Find det A, and prove that A is nonsingular.
6. Let $\mathbf{P}$ be the finite projective plane of Exercise 5, with incidence matrix A. For a collineation μ of $\mathbf{P}$ define *collineation* matrices $P = (p_{ij})$, $L = (l_{ij})$ as follows: $p_{ij} = 1$ if $P_i\mu = P_j$, $p_{ij} = 0$ if $P_i\mu \neq P_j$; $l_{ij} = 1$ if $l_i\mu = l_j$, $l_{ij} = 0$ if $l_i\mu \neq l_j$. Prove that P, L are nonsingular and that $PA = AL$.

2. DESARGUES' CONFIGURATION AND TRANSITIVITY

The algebraic system formed by the coordinates of an arbitrary projective plane $\mathbf{P}$, amounting to two weakly related loops, is very wild in comparison with the system formed by the coordinates of a $\mathbf{P}_2F$, that is, a field. It is the main goal of this chapter to show that the closer the geometric structure of $\mathbf{P}$ is to that of $\mathbf{P}_2F$, the closer will be the associated algebraic structures. In this section we shall examine the most important device for measuring closeness of the geometric structures of projective planes, namely, the *Desargues* configuration* [Figure III.2.1(a)].

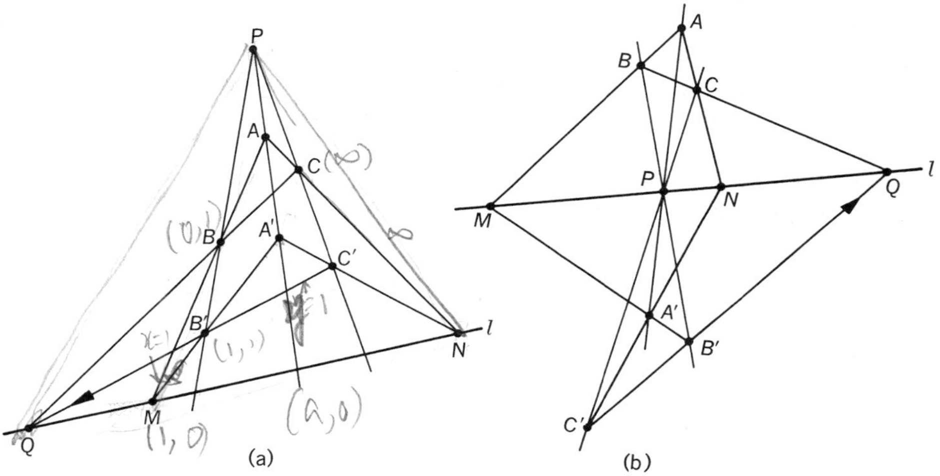

Fig. III.2.1

DEFINITION 1 Let P be a point, l a line of a projective plane $\mathbf{P}$. $\mathbf{P}$ is *(P, l)-Desarguesian* if, whenever A, B, C, A', B', C' are distinct points of $\mathbf{P}$ such that:

(1) P, A, A' are collinear, P, B, B' are collinear, and P, C, C' are collinear;

(2) $AB \neq A'B'$, $AC \neq A'C'$, and $BC \neq B'C'$;

(3) $AB \wedge A'B'$ and $AC \wedge A'C'$ are on l;

then $BC \wedge B'C'$ is on l.

The reader should check (Exercise 1) that the definition covers all nontrivial cases. Note that in the definition it is not specified whether P and l are incident [Figure III.2.1(b)] or nonincident [Figure III.2.1(a)].

We call $\mathbf{P}$ *Desarguesian* if it is (P, l)-Desarguesian for every P and l. Desargues' original theorem was essentially the real case of the following:

* Gérard Desargues, 1593–1662, was one of the best geometers of the century that saw Descartes, Newton, and Pascal.

THEOREM 1 $\mathbf{P}_2F$ is Desarguesian where F is any field.

PROOF. Given the setup of Definition 1, let $P:\mathbf{p}$, $A:\mathbf{a}$, $B:\mathbf{b}$, $C:\mathbf{c}$. By Theorem II.2.1, $A':\mathbf{p} + a'\mathbf{a}$, $B':\mathbf{p} + b'\mathbf{b}$, $C':\mathbf{p} + c'\mathbf{c}$ for some a', b', $c' \neq 0$ in F. Now $a'\mathbf{a} - b'\mathbf{b} = (\mathbf{p} + a'\mathbf{a}) - (\mathbf{p} + b'\mathbf{b})$ represents a point on both AB and $A'B'$ (note that this vector is not $\mathbf{0}$ because $A \neq B$). Hence, $AB \wedge A'B' : a'\mathbf{a} - b'\mathbf{b}$. Similarly, $AC \wedge A'C' : a'\mathbf{a} - c'\mathbf{c}$. But $(a'\mathbf{a} - c'\mathbf{c}) + (-1)(a'\mathbf{a} - b'\mathbf{b}) = b'\mathbf{b} - c'\mathbf{c}$. Hence, by Theorem II.2.1, $BC \wedge B'C'$ is on the line l through $AB \wedge A'B'$ and $AC \wedge A'C'$. ∎

Of course, not all projective planes are Desarguesian (Exercise 4).

We proceed now to an important connection between Definition 1 and the study of collineations (see Section II.5 if necessary).

DEFINITION 2 Let P be a point, l a line in a projective plane $\mathbf{P}$. $\mathbf{P}$ is *(P, l)-transitive* if, given any two points A, A' such that

(1) $P \neq A$ and $P \neq A'$,
(2) A is not on l and A' is not on l,
(3) P, A, A' are collinear,

there exists a (P, l)-perspectivity π of $\mathbf{P}$ such that $A\pi = A'$.

The conditions of the definition are necessary (Exercise 6). Thus, $\mathbf{P}$ being (P, l)-transitive means that $\mathbf{P}$ has all possible (P, l)-perspectivities (Exercise 7).

THEOREM 2 Let P be a point, l a line of a projective plane $\mathbf{P}$. $\mathbf{P}$ is (P, l)-Desarguesian if and only if $\mathbf{P}$ is (P, l)-transitive.

PROOF. Let $\mathbf{P}$ be (P, l)-transitive. Suppose we have the setup of Definition 1. Let $M = AB \wedge A'B'$, $N = AC \wedge A'C'$, $Q = BC \wedge B'C'$. We must show Q is on l. Let π be the (P, l)-perspectivity such that $A\pi = A'$. (Check that we can assume the conditions of Definition 2 hold!) Then $(AB)\pi = (AM)\pi = A'M = A'B'$, and $(PB)\pi = PB = PB'$. Hence, $B\pi = (AB \wedge PB)\pi = A'B' \wedge PB' = B'$. Similarly, $C\pi = C'$. Let $G = BC \wedge l$. Then, $G = G\pi$ is also on $(BC)\pi = B'C'$, so $G = BC \wedge B'C' = Q$. Thus, $Q = G$ is on l.

Conversely, let $\mathbf{P} = (\mathcal{P}, \mathcal{L}, \epsilon)$ be (P, l)-Desarguesian. Suppose we have the setup of Definition 2. We may assume $A \neq A'$. Let $a = AA'$.

Define a function π_A on $\mathcal{P} - \tilde{a}$ as follows: $X\pi_A = (AX \wedge l)A' \wedge PX$. Let B be any point not on a or l. Let $B' = B\pi$. Then $B \neq B'$, so we may let $b = BB'$. Define a function π_B on $\mathcal{P} - \tilde{b}$ as follows: $X\pi_B = (BX \wedge l)B' \wedge PX$.

ASSERTION. If X is not on a or b, then $X\pi_A = X\pi_B$. For, by definition $AB \wedge A'B'$ and $AX \wedge A'X\pi_A$ are on l, and so proving $X\pi_A = X\pi_B$ amounts to proving $BX \wedge B'X\pi_A$ is also on l (draw a sketch to see this). But this is true because $\mathbf{P}$ is (P, l)-Desarguesian.

Now define a function π on $\mathcal{P} \cup \mathcal{L}$ as follows: On $\mathcal{P}$: $X\pi = X\pi_A$ for X

not on a, $X\pi = X\pi_B$ for $X \neq P$ on a, and $P\pi = P$. On $\mathfrak{L}$: $h\pi = (h \wedge a)\pi(h \wedge l)$ if $l \wedge a \notin h$, $h\pi = (h \wedge b)\pi(h \wedge l)$ if $l \wedge a \in h$, $l \neq h$, and $l\pi = l$. It is immediately seen from this definition that $Q\pi = Q$ for all Q on l, that $k\pi = k$ for all k through P, and that $A\pi = A'$ (draw another sketch). Hence, if we show that π preserves incidence and is one–one and onto on $\mathcal{P}$, then by Theorem II.5.1, π will be the required perspectivity.

π preserves incidence: Suppose $Y \epsilon h$. If $Y \epsilon a$ or $Y \epsilon l$ then it is immediate from the definition that $Y\pi \epsilon h\pi$, no matter what h is. Now assume Y not on a or l. If $Y = B$, then $Y\pi \epsilon h\pi$ would again be immediate from the definition. But by the assertion (above), we may assume $Y = B$! (since $X\pi_Y = X\pi_A = X\pi_B$ whenever all three functions are defined).

π is one–one on points: Suppose $X\pi = Y\pi$. Then X and Y are on the same line, say h, through P. If $h \neq a$ then $X\pi = X\pi_A = (AX \wedge l)A' \wedge h = Y\pi = Y\pi_A = (AY \wedge l)A' \wedge h = Z$, say. Then $A'Z \wedge l = AX \wedge l = AY \wedge l = W$, say, and so $X = AW \wedge h = Y$. If $h = a$ but neither X nor Y is P, then $X\pi_B = Y\pi_B$ and, arguing just as before, $X = Y$. Finally, if X or Y is P, then $X = Y = P$.

π is onto $\mathcal{P}$: Let Y be any point. We must find a point X such that $X\pi = Y$. If $Y = P$, take $X = P$. Now assume $Y \neq P$. Let $h = YP$. If $h \neq a$, take $X = h \wedge A(A' \wedge l)$. Then $X \notin a$, so $X\pi = X\pi_A = \{A[h \wedge A(A'Y \wedge l)] \wedge l\}A' \wedge P[h \wedge A(A'Y \wedge l)] = (A'Y \wedge l)A' \wedge h = A'Y \wedge h = Y$. Finally, if $h = a$, take $X = a \wedge B(B'Y \wedge l)$. Then, just as before, $X\pi = X\pi_B = Y$. ∎

This theorem will be of considerable use to us in the next section.

A number of interesting questions naturally arise in the course of studying this and the previous sections. For example: If another selection of l_∞, g, h is made, what happens to (S, T)? Do some (nontrivial) cases of Desargues' configuration occur in every projective plane? We shall postpone discussion of these matters and proceed with the main business of this chapter.

Exercises

1. Prove that in every setup of Figure III.2.1 in which Definition 1 is not applicable (for instance, if A, B, C are collinear) the conclusion $AC \wedge A'C' \in l$ holds in every projective plane.
2. Prove directly (without Theorem 1) that the projective planes of orders 2 and 3 are Desarguesian.
3. Let D be a division ring (Definition I.2). Define $\mathbf{P}_2D$ like $\mathbf{P}_2F$, F a field, except that lines are classes of the form $\{\mathbf{u}t \mid t \neq 0\}$. (Then a point $\{t\mathbf{x} \mid t \neq 0\}$ is on this line iff $\mathbf{x} \cdot \mathbf{u} = 0$, not necessarily iff $\mathbf{u} \cdot \mathbf{x} = 0$.) Prove that $\mathbf{P}_2D$ is a Desarguesian projective plane.

*4. (The Moulton plane.) Modify the real projective plane as follows: Points are unchanged. The new line is unchanged. Ordinary vertical lines and ordinary lines with negative slope are unchanged. A line l with ordinary equation $y = mx + b$, $m > 0$, is replaced by a broken line:

$$y = \begin{cases} mx + b, & x \geq -bm^{-1} \\ 2(mx + b), & x < -bm^{-1}. \end{cases}$$

The new point on l is retained on the broken line. Prove that this modified system is a projective plane. Show, by drawing an explicit case, that it is not (P, l)-Desarguesian for all P and l.

5. (See Exercise 4.) Prove that the Moulton plane is $[(0), l_\infty]$-Desarguesian.
6. Let π be a (P, l)-perspectivity of a projective plane. Let A be a point that is moved by π. Prove: (1) $A \neq P$ and $A\pi \neq P$. (2) A is not on l and $A\pi$ is not on l. (3) P, A, $A\pi$ are collinear.
7. Let π_1 and π_2 be (P, l)-perspectivities of a projective plane. Let A be a point different from P and not on l. Prove that if $A\pi_1 = A\pi_2$, then $\pi_1 = \pi_2$.
8. State the dual of Definition 1. Compare this with the converse of Definition 1.
9. Prove that in a Desarguesian projective plane the dual of Definition 1 holds everywhere. (Thus Theorem II.3.4 holds in Desarguesian projective planes.)
10. (H. S. M. Coxeter.) Show that Figure III.2.1 consists, in five different ways, of a quadrangle and a quadrilateral so situated that the sides of the quadrangle pass through the vertices of the quadrilateral.

3. CONFIGURATION THEOREMS

Let **P** be a projective plane. Select l_∞, g, h and coordinatize **P**, as in Section 1. Obtain the planar ternary ring (S, T) with its binary operations $+$ and $\cdot$. We now begin the sequence of theorems showing how, the more cases of Desargues' configuration there are in **P**, the closer is $(S, +, \cdot)$ to a field.

DEFINITION 1 A planar ternary ring (S, T) is *linear* if for all a, b, c in S

$$\mathsf{T}(a, b, c) = ab + c.^*$$

The student may find it useful to make copies on separate sheets of Figures III.2.1(a) and (b). All theorems in this and the next section refer to these figures.

---

* As is customary in algebra, $ab + c$ means $(ab) + c$.

THEOREM 1 (S, T) is linear if and only if the Desargues, configuration holds in **P** whenever $l \colon l_\infty$, $P \colon (\infty)$, $Q \colon (0)$ and $AA' \colon [0]$.

PROOF. We may assume $A \colon (0, a)$, $A' \colon (0, a')$, $M \colon (m)$, $N \colon (n)$, $B \colon (u, b)$, $B' \colon (u, b')$, $C \colon (v, b)$, $C' \colon (v, c')$.

If (S, T) is linear, then $mu + b = \mathsf{T}(m, u, b) = a = \mathsf{T}(n, v, b) = nv + b$, so $mu + b = nv + b$. Then, since $(S, +)$ is a loop, $mu = nv$. In the same way, $mu + b' = a' = nv + c' = mu + c'$, and so $b' = c'$. Thus, $B' \colon (u, b')$, $C' \colon (v, b')$ are collinear with $Q \colon (0)$, as was to be proved.

If the given cases of Desargues' configuration are present, then $b' = c'$ for all choices of the above points. Choose $n = 1$; $b' = 0$; m, u, b arbitrary. Then $\mathsf{T}(m, u, b) = a = \mathsf{T}(1, v, b) = v + b = \mathsf{T}(1, v, 0) + b = a' + b = \mathsf{T}(m, u, 0) + b = mu + b$, that is, $\mathsf{T}(m, u, b) = mu + b$ for all m, u, b. ∎

THEOREM 2 (S, T) is linear and $(S, +)$ is associative if and only if **P** is $((\infty), l_\infty)$-Desarguesian.

PROOF. We may assume $Q \colon (q)$, $M \colon (m)$, $N \colon (n)$, $A \colon (u, a)$, $A' \colon (u, a')$, $B \colon (v, b)$, $B' \colon (v, b')$, $C \colon (w, c)$, $C' \colon (w, c')$. By Theorem 1 our hypotheses imply linearity. Thus, in either half of the proof we have:

$$mu + a = mv + b, \tag{1}$$

$$mu + a' = mv + b', \tag{2}$$

$$nu + a = nw + c, \tag{3}$$

$$nu + a' = nw + c', \tag{4}$$

$$qv + b = qw + c. \tag{5}$$

If $(S, +)$ is associative, then from (1) and (2)

$$-a + a' = -b + b', \tag{6}$$

and from (3) and (4)

$$-c + c' = -a + a'. \tag{7}$$

Then: $qw + c' \overset{(5)}{=} qv + b - c + c' \overset{(6,7)}{=} qv + b - b + b' = qv + b'$, that is, $qw + c' = qv + b'$. Hence, $Q \colon (q)$, $B' \colon (v, b')$, $C' \colon (w, c')$ are collinear, as desired.

Conversely, if **P** is $((\infty), l_\infty)$-Desarguesian, choose $q = v = a' = 0$; $n = 1$; w, b', a arbitrary. Then Eqs. (1)–(5) become:

$$mu + a = b, \tag{1'}$$

$$mu = b', \tag{2'}$$

$$u + a = w + c, \tag{3'}$$

$$u = w + c', \tag{4'}$$

$$b = c, \tag{5'}$$

and the conclusion of the Desarguesian configuration becomes

$$b' = c'. \tag{8}$$

Then, for all w, b', a: $(w + b') + a \overset{(8)}{=} (w + c') + a \overset{(4')}{=} u + a \overset{(3')}{=} w + c \overset{(5')}{=} w + b \overset{(1')}{=} w + (mu + a) \overset{(2')}{=} w + (b' + a)$, that is, $(S, +)$ is associative. ∎

THEOREM 3 $\mathbf{P}$ is $((0), l_\infty)$-Desarguesian if, and only if, for all m, a, x, y;

$$\mathsf{T}(m, a + x, y) = \mathsf{T}(m, a, \mathsf{T}(m, x, y)). \tag{*}$$

PROOF. Suppose $\mathbf{P}$ is $((0), l_\infty)$-Desarguesian. Let $a \in S$. By Theorem 2.2 there is a $((0), l_\infty)$-perspectivity ϕ_a of $\mathbf{P}$ such that $(0, 0)\phi_a = (a, 0)$. For any (x, y), $(x, y)\phi_a$ is of the form (x', y). This determines a function from x to x', which we call θ_a. [θ_a does not depend on y, because ϕ_a fixes (∞).] Thus, $(x, y)\phi_a = (x\theta_a, y)$. For any $[m, k]$, $[m, k]\phi_a$ is of the form $[m, n]$. Now, $(0, k) \in [m, k]$, hence, $(0, k)\phi_a = (a, k) \in [m, n]$, and so $\mathsf{T}(m, a, k) = n$. Thus, $[m, k]\phi_a = [m, \mathsf{T}(m, a, k)]$. Now, $(x, y) \in [m, k] \rightarrow (x, y)\phi_a \in [m, k]\phi_a$, that is, $\mathsf{T}(m, x, y) = k \rightarrow \mathsf{T}(m, x\theta_a, y) = \mathsf{T}(m, a, k)$. Hence, $\mathsf{T}(m, x\theta_a, y) = \mathsf{T}(m, a, \mathsf{T}(m, x, y))$ for all m, a, x, y. Setting $m = 1$, $y = 0$, this becomes $\mathsf{T}(1, x\theta_a, 0) = x\theta_a = \mathsf{T}(1, a, \mathsf{T}(1, x, 0)) = \mathsf{T}(1, a, x) = a + x$, and so $x\theta_a = a + x$ for all a, x. Substituting in the previous equation, we have (*).

Conversely, suppose (*). Let $t \in S$. Consider the function ϕ_t defined on the points and lines of $\mathbf{P}$ as follows:

$$\begin{aligned}
(x, y)\phi_t &= (t + x, y),\\
(m)\phi_t &= (m),\\
(\infty)\phi_t &= (\infty),\\
[m, k]\phi_t &= [m, \mathsf{T}(m, t, k)],\\
[k]\phi_t &= [t + k],\\
l_\infty\phi_t &= l_\infty.
\end{aligned}$$

We check that ϕ_t is a collineation. ϕ_t is clearly one–one and onto on points, so if we can show that ϕ_t preserves incidence, it will follow that ϕ_t is one–one and onto on lines and is a collineation. All cases of incidence preservation are obvious except $(x, y) \in [m, k] \rightarrow (x, y)\phi_t \in [m, k]\phi_t$. To check this case, suppose $\mathsf{T}(m, x, y) = k$. Then, $\mathsf{T}(m, t + x, y) \overset{(*)}{=} \mathsf{T}(m, t, \mathsf{T}(m, x, y)) = \mathsf{T}(m, t, k)$, and so $(x, y)\phi_t \in [m, k]\phi_t$.

Now any two points not on l_∞ that are collinear with (0) can be assumed to have coordinates (b, c) and (b', c). Then $\phi_{b'-b}$ is a $((0), l_\infty)$-perspectivity sending (b, c) to (b', c). Thus, $\mathbf{P}$ is $((0), l_\infty)$-transitive and, by Theorem 2.2, $((0), l_\infty)$-Desarguesian. ∎

We say $(S, +, \cdot)$ is *left distributive* if $a(b + c) = ab + ac$ for all a, b, c, and *right distributive* if $(a + b)c = ac + bc$ for all a, b, c.

COROLLARY (*) in the above theorem implies $(S, +)$ is associative. If (S, T) is linear, then (*) implies $(S, +, \cdot)$ is left distributive.

PROOF. In (*) set $m = 1$ to obtain $(a + x) + y = a + (x + y)$ for all a, x, y. If (S, T) is linear, set $y = 0$ in (*) to obtain $m(a + x) = ma + mx$ for all m, a, x. ∎

THEOREM 4 Suppose (S, T) is linear and $(S, +)$ is associative. Then $(S, +, \cdot)$ is right distributive if and only if $\mathbf{P}$ is $((\infty), [0])$-Desarguesian.

PROOF. Suppose $\mathbf{P}$ is $((\infty), [0])$-Desarguesian. Let $a \in S$. By Theorem 2.2 there is an $((\infty), [0])$-perspectivity ϕ_a such that $(0)\phi_a = (a)$. Now, (x, y) is of the form (x, y') where, this time, y' depends on x and y. In other words, a function $\psi_a : S^2 \to S$ is determined by $(x, y)\phi_a = (x, (x, y)\psi_a)$. Also, a function $\theta_a : S \to S$ is determined by $(m)\phi_a = (m\theta_a)$. Then, $[m, k]\phi_a = [m\theta_a, k]$, so $(x, y) \in [m, k]$ iff $(x, (x, y)\psi_a) \in [m\theta_a, k]$, that is, $mx + y = k$ iff $m\theta_a \cdot x + (x, y)\psi_a = k$. This is equivalent to the equation

$$mx + y = m\theta_a \cdot x + (x, y)\psi_a, \tag{1}$$

for all m, a, x, y. Setting $m = 0 : y = ax + (x, y)\psi_a$. Then (1) becomes $m\theta_a \cdot x - ax + y = mx + y$, whence

$$m\theta_a \cdot x = mx + ax. \tag{2}$$

Setting $x = 1 : m\theta_a = m + a$. Then (2) becomes $(m + a)x = mx + ax$ for all m, a, x, which is right distributivity.

Conversely, suppose (S, T) is linear, $(S, +)$ is associative, and $(S, +, \cdot)$ is right distributive. For each t in S define

$$\begin{aligned}
(x, y)\phi_t &= (x, -tx + y),\\
(m)\phi_t &= (m + t),\\
(\infty)\phi_t &= (\infty),\\
[m, k]\phi_t &= [m + t, k],\\
[k]\phi_t &= [k],\\
l_\infty\phi_t &= l_\infty.
\end{aligned}$$

This time two things are not completely obvious in the check that ϕ_t is a collineation, and, hence, an $((\infty), [0])$-perspectivity. First, ϕ_t is onto on points of the form $(x, y) : (x, tx + y)\phi_t = (x, -tx + tx + y) = (x, y)$. Second, $(x, y) \in [m, k] \to (x, y)\phi_t \in [m, k]\phi_t$: if $mx + y = k$ then $(m + t)x + (-tx + y) = mx + tx - tx + y = mx + y = k$.

Now any two points not on [0] or l_∞ that are collinear with (∞) may be assumed to have coordinates (b, c) and (b, c'), $b \neq 0$. The existence of t in S such that $(b, c)\phi_t = (b, c')$ is equivalent to the existence of t in S such that $tb = c - c'$. If $c = c'$, take $t = 0$. If $c \neq c'$, the existence (and uniqueness) of t follows from the fact that $(S - \{0\}, \cdot)$ is a loop. If the points are on l_∞, they may be assumed to have coordinates (m) and (n). Then, $(m)\phi_{-m+n} = (n)$. Thus, **P** is $((\infty), [0])$-transitive and so, by Theorem 2.2, $((\infty), [0])$-Desarguesian. ∎

Note that so far we have only used cases of the Desargues configuration in which P is on l [Figure III.2.1(b)]. This special configuration is often called the *little Desargues configuration*. We cannot get away with this special case in the following theorem.

THEOREM 5 (S, T) is linear and $(S, \cdot)$ is associative if and only if **P** is $((0), [0])$-Desarguesian.

PROOF. Suppose that **P** is $((0), [0])$-Desarguesian. Let $a \in S$. We may assume $a \neq 0$. By Theorem 2.2 there is a $((0), [0])$-perspectivity ϕ_a of **P** such that $(1, 0)\phi_a = (a, 0)$. Define functions ψ_a, θ_a from S to S by: $(x, y)\phi_a = (x\psi_a, y)$, $(m)\phi_a = (m\theta_a)$. Then, $[m, k]\phi_a = [m\theta_a, k]$ and $[k]\phi_a = [k\psi_a]$. Note that $1\psi_a = a$. Now, $\mathsf{T}(m, x, y) = k$ iff $\mathsf{T}(m\theta_a, x\psi_a, y) = k$. This is equivalent to

$$\mathsf{T}(m, x, y) = \mathsf{T}(m\theta_a, x\psi_a, y). \tag{1}$$

Setting $y = 0$:

$$mx = m\theta_a \cdot x\psi_a. \tag{2}$$

Setting $x = 1$ in (2):

$$m = m\theta_a \cdot a. \tag{3}$$

Setting $m = a$ in (3): $a = a\theta_a \cdot a$. Then, since $(S - \{0\}, \cdot)$ is a loop,

$$a\theta_a = 1. \tag{4}$$

Replace m in (3) by ma and use the loop property to obtain

$$m = (ma)\theta_a. \tag{5}$$

Then $ax \overset{(2)}{=} a\theta_a \cdot x\psi_a \overset{(4)}{=} x\psi_a$, that is,

$$x\psi_a = ax. \tag{6}$$

Then (2) becomes

$$mx = m\theta_a(ax). \tag{7}$$

Replace m in (7) by ma: $(ma)x = (ma)\theta_a(ax) \overset{(5)}{=} m(ax)$ for all m, a, x. Thus, $(S, \cdot)$ is associative. Now in (1) set $m = a$ and use (4) and (6) to obtain: $\mathsf{T}(a, x, y) = \mathsf{T}(1, ax, y) = ax + y$. Hence, (S, T) is linear.

Conversely, suppose that (S, T) is linear and has associative multiplication. For t in $S - \{0\}$, define ϕ_t as follows:

$$\begin{aligned}(x, y)\phi_t &= (tx, y),\\ (m)\phi_t &= (mt^{-1}),^*\\ (\infty)\phi_t &= (\infty),\\ [m, k]\phi_t &= [mt^{-1}, k],\\ [k]\phi_t &= [tk],\\ l_\infty\phi_t &= l_\infty.\end{aligned}$$

If $mx + y = k$, then $(mt^{-1})(tx) + y = mx + y = k$, so $(x, y) \in [m, k] \rightarrow (x, y)\phi_t \in [m, k]\phi_t$. All other conditions that ϕ_t be a ((0), [0])-perspectivity are clear. Two points not on [0] or l_∞ that are collinear with (0) can be assumed to have coordinates of the form (b, c), (b', c), $b \neq 0 \neq b'$. Then, $(b, c)\phi_{b'b^{-1}} = (b', c)$. Points on l_∞ may be assumed to have coordinates (m), (n). Then $(m)\phi_{n^{-1}m} = (n)$. Thus, **P** is ((0), [0])-transitive and, by Theorem 2.2, ((0), [0])-Desarguesian. ∎

We may summarize the theorems of this section by saying that (S, T) is linear and $(S, +, \cdot)$ is a ring if and only if **P** is $((\infty), l_\infty)$-, $((0), l_\infty)$-, $((\infty), [0])$-, and ((0), [0])-Desarguesian. However, as we shall see in the next section, a stronger statement holds.

Exercises

1. Using the usual ruler and compass construction for addition in the Euclidean plane ([8], vol. II, pp. 30–31), draw the figure for the associative law of addition in $\mathbf{P}_2\mathrm{R}$. Use this to interpret Theorem 2.
2. Do Exercise 1 for multiplication and Theorem 5.
3. Having done Exercises 1 and 2, try to interpret Theorems 3 and 4 in $\mathbf{P}_2\mathrm{R}$.
4. Prove that if a projective plane **P** is (P, l)-Desarguesian and if μ is any collineation of **P**, then **P** is also $(P\mu, l\mu)$-Desarguesian. (See Exercise II.5.2.)
5. (We refer to Figure III.2.1.) Let (S, T) be a planar ternary ring of a coordinatized projective plane. Prove that $\mathsf{T}(a, 1, b) = a + b$ for all a and b if and only if Desargues' configuration holds whenever $l \colon l_\infty$, $P \colon (\infty)$, $Q \colon (0)$, $AA' \colon [0]$, $B \colon (1, 0)$, and $N \colon (1)$.

* $(S - \{0\}, \cdot)$ is now a group; so t^{-1} has the usual meaning.

4. CONFIGURATION THEOREMS (CONTINUED)

In this section we obtain some algebraic results that we shall use to extend and interpret the theorems of Section 3.

DEFINITION 1 A *left planar V–W system** is a linear planar ternary ring with associative addition and the left distributive law:

$$a(b + c) = ab + ac \text{ for all } a, b, \text{ and } c.$$

Before discussing the geometric significance of these systems, we shall study them as algebraic systems.

THEOREM 1 Let $(S, +, \cdot)$ be a left planar V–W system.

(1) If $a \neq 1$, then for each b in S there is a unique x such that

$$ax + b = x.$$

(2) Addition in S is commutative.

PROOF

(1) Since $(S, +, \cdot)$ is a linear planar ternary ring, the equations $(a \neq 1)$

$$au + v = b, \qquad u + v = 0$$

have a unique solution $u = u_0$, $v = v_0$. Then $u_0 = -v_0$ and $a(-v_0) + v_0 = b$. Now $a(-v_0) + av_0 = a(-v_0 + v_0) = a \cdot 0 = 0$, so $a(-v_0) = -(av_0)$. Thus, $av_0 + b = v_0$, that is, v_0 is a solution of $ax + b = x$. If v_1 is also a solution, $av_1 + b = v_1$, then $v_0 - v_1 = av_0 + b - b - av_1 = av_0 - av_1 = a(v_0 - v_1)$. Then, if $v_0 \neq v_1$, we would have $0 \neq v_0 - v_1 = a(v_0 - v_1)$, whence $a = 1$, a contradiction. Hence, $v_0 = v_1$, proving uniqueness.

(2) If $a + b \neq b + a$ for some a, b in S, then $b + a - b \neq a$. Obviously, $a \neq 0$, so we may write $b + a - b = ra$ for some $r \neq 1$. By (1) there is a unique x_0 in S such that $rx_0 + b = x_0$. But $-[r(x_0 + a)] + (x_0 + a) = -(rx_0 + ra) + x_0 + a = -(ra) - (rx_0) + x_0 + a = -(ra) + b + a = -(b + a - b) + b + a = b - a - b + b + a = b$, that is, $r(x_0 + a) + b = x_0 + a$. Thus, both x_0 and $x_0 + a$ are the unique solution of $rx + b = x$, so $x_0 = x_0 + a$. But then $a = 0$, a contradiction. ∎

This theorem yields a nice geometrical result.

DEFINITION 2 Let P be a point (l a line) of a projective plane **P**. **P** is *P-Desarguesian* (*l-Desarguesian*) if it is (P, x)-Desarguesian for every line x through P [respectively, (X, l)-Desarguesian for every point X on l].

---

* Named for Oswald Veblen and J. H. Maclagan-Wedderburn.

THEOREM 2 Let A and B be distinct points in a projective plane **P**. If **P** is (A, AB)-Desarguesian and (B, AB)-Desarguesian, then **P** is AB-Desarguesian.

PROOF. Coordinatize **P** in such a way that $A\colon(0)$, $B\colon(\infty)$. Then, $AB\colon l_\infty$ and by hypothesis **P** is $((0), l_\infty)$-Desarguesian and $((\infty), l_\infty)$-Desarguesian. By Theorem 3.2 the resulting planar ternary ring $(S, \mathsf{T}) = (S, +, \cdot)$ is linear and has associative addition. By Theorem 3.3, $(S, +, \cdot)$ is left distributive. Hence, $(S, +, \cdot)$ is a left planar V–W system.

Now let P be any point different from A and B on AB. $P\colon(p)$ for some p in S. Let U and U' be points not on $AB = l_\infty$ that are collinear with P. Then, $U\colon(u_1, u_2)$ and $U'\colon(u_1', u_2')$ where

$$pu_1 + u_2 = pu_1' + u_2'. \tag{*}$$

Let $a = u_1' - u_1$, $b = -u_2 + u_2'$. Define a function ϕ as follows:

$$(x, y)\phi = (a + x, y + b),$$

$$(m)\phi = (m),$$

$$(\infty)\phi = (\infty),$$

$$[m, k]\phi = [m, ma + k + b],$$

$$[k]\phi = [a + k],$$

$$l_\infty\phi = l_\infty.$$

If $mx + y = k$, then $m(a + x) + y + b = ma + mx + y + b = ma + k + b$. Thus, $(x, y) \in [m, k] \rightarrow (x, y)\phi \in [m, k]\phi$. With this it becomes obvious that ϕ is a perspectivity with axis l_∞ and that $(u_1, u_2)\phi = (u_1', u_2')$. But it is not obvious that ϕ has center P! We must still show that ϕ fixes every line through P, that is, that $[p, k]\phi = [p, k]$ for all k in S, that is, that $pa + k + b = k$ for all k in S. Now, by Theorem 1(2), addition is commutative. Then, $pa + k + b = p(u_1' - u_1) + k - u_2 + u_2' = k + (pu_1' + u_2') - (pu_1 + u_2) \overset{*}{=} k$. ∎

We can now recast the results of the previous section.

DEFINITION 3 A *left planar near-field* is a left planar V–W system with associative multiplication.

THEOREM 3 Let **P** be a projective plane. Let (S, T) be the planar ternary ring obtained from some coordinatization of **P**. Then:

(1) (S, T) is a left planar V–W system if and only if **P** is l_∞-Desarguesian.
(2) (S, T) is a left planar near-field if and only if **P** is both l_∞-Desarguesian and $((0), [0])$-Desarguesian.

PROOF

(1) If (S, T) is a left planar V–W system, then by Theorem 3.2 **P** is $((\infty), l_\infty)$-Desarguesian, and then by Theorem 3.3 **P** is $((0), l_\infty)$-Desarguesian. Hence, by Theorem 2 **P** is l_∞-Desarguesian.

Conversely, if **P** is l_∞-Desarguesian, then it is in particular $((\infty), l_\infty)$- and $((0), l_\infty)$-Desarguesian. Hence, by Theorems 3.2 and 3.3, (S, T) is a left planar V–W system.

(2) If (S, T) is a left planar near-field, it is in particular a left planar V–W system, so by (1) **P** is l_∞-Desarguesian. Besides, multiplication is associative, so by Theorem 3.5 **P** is also $((0), [0])$-Desarguesian.

Conversely, if **P** is l_∞- and $((0), [0])$-Desarguesian, then by (1) and Theorem 3.5 (S, T) is a left planar near-field. ▮

We next show just how far one can go using only the little Desargues configuration.

DEFINITION 4 A projective plane **P** is *Moufang** if it is l-Desarguesian for every line l in **P**.

LEMMA In a Moufang plane, Desargues' configuration holds whenever $C' \in AB$. (Notation as in Figure III.2.1. The reader should make a drawing of this case for use below.)

PROOF. Consider triangles NCQ and $A'PB'$. The triples A', N, C'; P, C, C'; B', Q, C'; and $NC \wedge A'P = A$, $CQ \wedge PB' = B$, C' are each collinear. Then, since the plane is (C', AB)-Desarguesian, $NQ \wedge A'B' = M' \in AB$. Then, $M = AB \wedge A'B' = M'$, so $M \in NQ$. ▮

Before the next theorem we need a little more algebra. If a planar ternary ring (S, T) is a left planar V–W system, then in order for $(S, +, \cdot)$ to be an alternative division ring (Section I.6) it suffices to have: (1) for all a, b, c, $(a + b)c = ac + bc$; (2) if $b \neq 0$ and $bb^* = 1$, then for all a, $b^*(ba) = a$; (3) for all $b \neq 0$ and all a, $(ab)b^{-1} = a$. In (2) set $a = 1$ to obtain $b^*b = 1$ $(= bb^*)$. Hence, after (2) has been proved we are justified in writing b^{-1} for b^*, as we do in (3). After (1) has been proved, we have the familiar laws $x0 = 0x = 0$ for all x, and $x(-y) = (-x)y = -(xy)$ for all x and y.

THEOREM 4 Let **P** be a projective plane and let (S, T) be the planar ternary ring obtained from some coordinatization of **P**. Then, (S, T) is linear and $(S, +, \cdot)$ is an alternative division ring if and only if **P** is Moufang.

PROOF. Suppose **P** is Moufang. Then **P** is l_∞- and $((\infty), [0])$-Desarguesian, so by Theorems 3(1) and 3.4 (S, T) is linear and $(S, +, \cdot)$ is a right distributive left planar V–W system. It remains to prove (2) and (3) above.

---

* Named for Ruth Moufang, who studied these planes in the 1930s.

(2): Suppose $b \neq 0$, $bb^* = 1$, a arbitrary. Let $P{:}(\infty)$, $A{:}(1, 1)$, $B{:}(ba, ba)$, $C{:}(b^*, 1)$, $A'{:}(1, b^*)$, $B'{:}(ba, b^*(ba))$, $C'{:}(b^*, b^*)$. The reader may check that these points satisfy the conditions of the lemma and that $M{:}(0, 0)$, $N{:}(0)$, and so $MN{:}[0, 0]$. Then, by the lemma, $Q_0 = (b^*, b^*)(ba, b^*(ba)) \wedge (b^*, 1)(ba, ba) \in [0, 0]$. Now let $P{:}(\infty)$, $A{:}(b^*, b^*)$, $B{:}(a, a)$, $C{:}(ba, b^*(ba))$, $A'{:}(b^*, 1)$, $B'{:}(a, ba)$, $C'{:}(ba, ba)$. The reader may check that the conditions of the lemma hold, and, hence, that $M{:}(0, 0)$, $N = Q_0$, and $Q = BC \wedge B'C' = BC \wedge [0, ba]$ lie on $MN{:}[0, 0]$. But then $Q{:}(0)$ and so, since B and C are collinear with Q, B and C have the same second coordinate, that is, $b^*(ba) = a$.

(3): Suppose $b \neq 0$, a arbitrary. Let $P{:}(0, 0)$, $A{:}(1, b^{-1})$, $B{:}(1, a)$, $C{:}(b^{-1}, b^{-1})$, $A'{:}(b, 1)$, $B'{:}(b, ab)$, $C'{:}(1, 1)$. Then $M{:}(\infty)$, $N{:}(0)$, the lemma applies, and so $Q_0 = (1, 1)(b, ab) \wedge (b^{-1}, b^{-1})(1, a) \in l_\infty$. Now, let $P{:}(0, 0)$, $A{:}(1, 1)$, $B{:}(1, ab)$, $C{:}(b, ab)$, $A'{:}(b^{-1}, b^{-1})$, $B'{:}(b^{-1}, (ab)b^{-1})$, $C'{:}(1, a)$. Then, $M{:}(\infty)$, $N = Q_0$, the lemma applies, and so $Q = BC \wedge B'C' = [0, ab] \wedge B'C'$ lies on $MN{:}l_\infty$. But then $Q{:}(0)$, and so B' and C' have the same second coordinate, that is, $(ab)b^{-1} = a$.

Conversely, suppose (S, T) is linear and $(S, +, \cdot)$ is an alternative division ring. By Theorem 3(1) **P** is l_∞-Desarguesian. Now, the mapping μ_1 defined by:

$$\begin{aligned}
(0, y)\mu_1 &= (y),\\
(x, y)\mu_1 &= (x^{-1}, -yx^{-1}) \text{ for } x \neq 0,\\
(\infty)\mu_1 &= (\infty),\\
(m)\mu_1 &= (0, m),\\
[m, k]\mu_1 &= [k, m],\\
[0]\mu_1 &= l_\infty,\\
[k]\mu_1 &= [k^{-1}] \text{ for } k \neq 0,\\
l_\infty\mu_1 &= [0],
\end{aligned}$$

is a collineation of **P**. (For example, suppose $(x, y) \in [m, k]$, $x \neq 0$. Then, $kx^{-1} - yx^{-1} = (k - y)x^{-1} = (mx)x^{-1} = m$, so $(x, y)\mu_1 \in [m, k]\mu_1$.) Then, by Exercise 4, **P** is also [0]-Desarguesian. Now, the mapping μ_2 defined by

$$\begin{aligned}
(x, y)\mu_2 &= (y, x),\\
(\infty)\mu_2 &= (0),\\
(0)\mu_2 &= (\infty),\\
(m)\mu_2 &= (m^{-1}) \text{ for } m \neq 0,\\
[0, k]\mu_2 &= [k],\\
[m, k]\mu_2 &= [m^{-1}, m^{-1}k] \text{ for } m \neq 0,\\
[k]\mu_2 &= [0, k],\\
l_\infty\mu_2 &= l_\infty,
\end{aligned}$$

is also a collineation of **P**, so again by Exercise 4 **P** is [0, 0]-Desarguesian. Finally, by Exercise 5, **P** is Moufang. ∎

At last we use the full strength of Desargues' configuration to obtain the following theorem.

THEOREM 5 Let **P** be a projective plane and let (S, T) be the planar ternary ring obtained from some coordinatization of **P**. Then (S, T) is linear and $(S, +, \cdot)$ is a division ring if and only if **P** is Desarguesian.

PROOF. If **P** is Desarguesian, then by Theorems 3.2, 3.3, 3.4, and 3.5 (S, T) is linear and $(S, +, \cdot)$ is a division ring. Conversely, if (S, T) is linear and $(S, +, \cdot)$ is a division ring, then by Theorems 3.5 and 4 **P** is a ((0), [0])-Desarguesian Moufang plane. Then, by Exercise 11, **P** is Desarguesian. ∎

Since there are division rings, such as the ring of quaternions, that are not fields, we now see that Desargues' configuration is not enough to complete our sequence of theorems leading to $\mathbf{P}_2F$ with F a field. We shall therefore introduce another configuration in the next section.

Exercises

1. Prove that a finite planar left V–W system is of prime-power order.
2. The dual of Theorem 2 must also be true. State it.

*3. (J. L. Zemmer.) (For readers with enough background in algebra.) A more direct approach to the systems of this section proceeds as follows. A *left V–W system* is a set S together with two binary operations $+$ and $\cdot$ satisfying:

$$(S, +) \text{ is a group;}$$
$$(S - \{0\}, \cdot) \text{ is a loop;}$$
$$a(b + c) = ab + ac.$$

A *left near-field* is a left V–W system with associative multiplication. A left V–W system or a left near-field is called *planar* if given any $a \neq 1$ and any b there exists a unique x such that $ax + b = x$.

Now show that the system $(S, +, \circ)$ defined below is an example of a nonplanar left near-field. (The existence of such was unknown for some time.)

Let R be the reals. Let t be an indeterminate. Let $S = \mathrm{R}(t)$. Define $+$ as in $\mathrm{R}(t)$. For $\alpha = p/q$ in S (p, q polynomials in t) define $\delta(\alpha) =$ degree p − degree q. Define $T : S \to S$ as follows: for $\alpha = \alpha(t)$ in S, $\alpha T = \alpha(t + 1)$. Now for α, β in S define $\alpha \circ \beta = \alpha(\beta T^{\delta(\alpha)})$. (Hint: $t \circ \xi + t = \xi$ has no solution ξ.)

4. Prove, by use of Exercise 3.10, that if a projective plane **P** is l-Desarguesian (P-Desarguesian) for some line l (point P) and if μ is an arbitrary collineation of **P**, then **P** is $l\mu$-Desarguesian ($P\mu$-Desarguesian).
5. Prove that if a projective plane **P** is l-Desarguesian for three non-concurrent lines l, then **P** is Moufang.
6. Every alternative division ring satisfies the *alternative associativity* laws $a(ab) = (aa)b$ and $(ab)b = a(bb)$ for all a and b. The following is a sketch of a proof of the first law. Fill in the details and give a proof of the second law.

$$(a + 1)\{[a^{-1} - (a + 1)^{-1}](ax)\} = x,$$

hence

$$ax = [a^{-1} - (a + 1)^{-1}]^{-1}[(a + 1)^{-1}x]. \qquad (*)$$

Set $x = a + 1$ in (*) to obtain (**). Set $x = ab + b$ in (*) and use (**) to prove $a(ab) = (aa)b$.
7. Prove the following generalization of Theorem 2, due to R. Baer. If a projective plane **P** is (A, l)- and (B, l)-Desarguesian, $A \neq B$, then **P** is (P, l)-Desarguesian for every P on AB. (Try a synthetic proof.)

*8. A projective plane is *A–B-Desarguesian*, where A and B are points, if it is (A, l)-Desarguesian for every line l through B. Prove that a projective plane is A–B-Desarguesian iff it is B–A-Desarguesian. (Hint: Let $A{:}(0)$, $B{:}(\infty)$.)

9. A projective plane is *l–m-Desarguesian*, where l and m are lines, if it is (P, m)-Desarguesian for every point P on l. Suppose a projective plane is A–B-Desarguesian, where $A \neq B$. Using Exercise 8, prove the plane is AB–AB-Desarguesian.

*10. Let A and B be points, a and b lines of a projective plane **P**. Suppose that **P** is (A, a)- and (B, b)-Desarguesian and that either (1) $A \notin a$, $A \notin b$, $B \in a$, and $B \notin b$, or (2) $A \in a$, $A \notin b$, $B \notin a$, and $B \notin b$. Prove that **P** is Desarguesian.

11. Prove that if a Moufang plane **P** is (P, l)-Desarguesian for some point P and line l not through P, then **P** is Desarguesian.

5. PAPPUS' CONFIGURATION AND THE THEOREM OF HESSENBERG

As we have said, another configuration must be introduced to complete the sequence of configuration theorems. We shall employ the configuration of Pappus* (Figure III.5.1).

---

* Pappus of Alexandria (probably fourth century A.D.) was the last great Greek geometer. Several hundred pages of his work have survived [7].

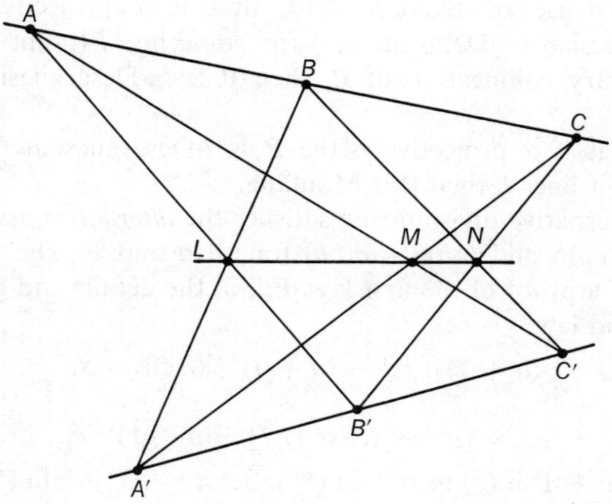

Fig. III.5.1

DEFINITION 1 A projective plane **P** is *Pappian* if whenever A, B, C; A', B', C' are distinct points of **P** such that

(1) A, B, C are collinear,
(2) A', B', C' are collinear,
(3) A, B, C are not on $A'B'$,
(4) A', B', C' are not on AB,

then $L = AB' \wedge A'B$, $M = AC' \wedge A'C$, $N = BC' \wedge B'C$ are collinear.

The reader should check (Exercise 1) that Definition 1 covers all nontrivial cases. In contrast to our use of Desargues' configuration, we shall not consider cases in which Pappus' configuration holds in only a part of the plane. We mention only one interesting result in this vein, due to R. P. Burn (to appear in *Mathematische Zeitschrift*). Let l and m be distinct lines of a projective plane **P**. We say **P** is *l–m-Pappian* if Definition 1 holds whenever A, B, C are on l and A', B', C' are on m. If **P** is l–m-Pappian for some l and m, then **P** is Pappian.

Our first theorem expresses one of the most important properties of Pappus' configuration; it was discovered by G. Hessenberg in 1905. Hessenberg's proof, however, covered only Subcase 1.1 (see below and Exercise 2). The full proof was given much later by A. Cronheim and D. Pedoe. It is, unfortunately, quite tedious!

THEOREM 1 Every Pappian projective plane is Desarguesian.

PROOF. We shall refer to Figure III.5.2. We write (XYZ) to indicate that points X, Y, Z are collinear, and $\sim(XYZ)$ to indicate that they are not

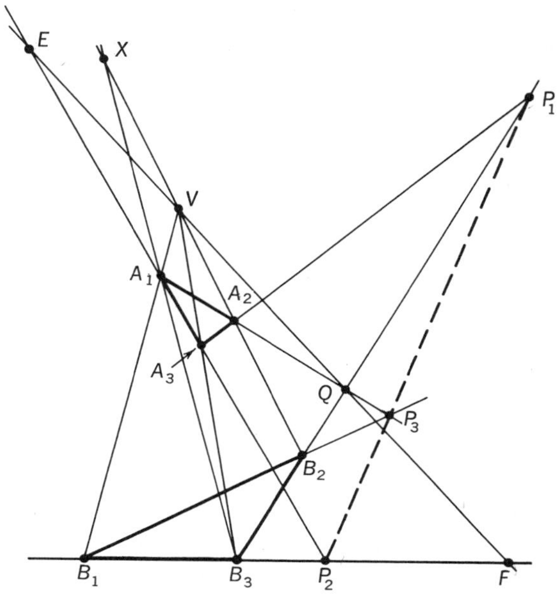

Fig. III.5.2

collinear. We are given (VA_iB_i), $i = 1, 2, 3$, and are to prove, using the Pappian property, $(P_1P_2P_3)$. Recall that we can assume: V, A_1, A_2, A_3, B_1, B_2, B_3 distinct, $\sim(A_1A_2A_3)$, and $\sim(B_1B_2B_3)$. We require two lemmas.

LEMMA I $\sim(A_iB_iB_k)$ and $\sim(A_iA_kB_k)$ for all $i \neq k$. This is immediate from our assumptions.

LEMMA II If for every rearrangement (i, j, k) of $(1, 2, 3)$ either $(A_iB_jB_k)$ or $(B_kA_iA_j)$, then either $(A_iB_jB_k)$ for all i, j, k or $(B_iA_jA_k)$ for all i, j, k.

Suppose, without loss of generality, that $\sim(B_1A_2A_3)$. Then, by hypothesis, $(A_2B_1B_3)$, and, since $\sim(B_1A_3A_2)$, $(A_3B_1B_2)$. Now, if $(B_3A_1A_2)$, then, since $(A_2B_1B_3)$, we would have $(A_1A_2B_1)$, contradicting Lemma I. Hence, $\sim(B_3A_1A_2)$, and so, by hypothesis, $(A_1B_2B_3)$. We have shown $(A_iB_jB_k)$ for all i, j, k, as required.

We shall now prove the theorem.

CASE 1. There is a rearrangement (i, j, k) of $(1, 2, 3)$ such that $\sim(A_iB_jB_k)$ and $\sim(B_kA_iA_j)$. We may assume $\sim(A_1B_2B_3)$ and $\sim(B_3A_1A_2)$. Let $Q = A_1A_2 \wedge B_2B_3$. Then: $\sim(A_1B_2B_3) \to Q \neq A_1$, $\sim(A_2B_2B_3) \to Q \neq A_2$, $\sim(A_3B_2B_3) \to Q \neq A_3$, $\sim(B_1B_2B_3) \to Q \neq B_1$, $\sim(A_1B_2A_2) \to Q \neq B_2$, $\sim(B_3A_2A_1) \to Q \neq B_3$, and $\sim(A_1A_2A_3) \to Q \neq V$.

ASSERTION. Definition 1 applies to A_3, B_3, V; Q, A_2, A_1. We must show that (3) and (4) hold. Supposing that the contrary holds, will lead to one of the following conclusions: $(A_1A_2A_3)$, $(A_1A_2B_3)$, (A_1A_2V), $(A_1A_3B_3)$, $(A_2A_3B_3)$, (QA_3B_3). The first three conclusions contradict our assumptions; the next two contradict Lemma II. Finally, (QA_3B_3), together with (QB_2B_3), implies $Q = B_3$, which is false.

Then, by Definition 1, (XEP_1), where $X = VA_2 \wedge A_1B_3$, $E = VQ \wedge A_1A_3$.

ASSERTION. Definition 1 applies to $B_1, A_1, V; Q, B_2, B_3$. Failure of (3) or (4) would imply $(B_1B_2B_3)$, $(A_1B_2B_3)$, (VB_2B_3), $(B_2B_1A_1)$, $(B_3B_1A_1)$, or (QB_1A_1). The first three are excluded by assumption; the next two by Lemma II. (QB_1A_1) implies (A_1A_2V), contrary to our assumption.

Then, (XFP_3), where $F = VQ \wedge B_1B_3$.

Next we *attempt* to check the hypotheses of Definition 1 for $F, Q, E; A_1, X, B_3$. $\sim(VA_1B_2) \rightarrow X \neq A_1$, $\sim(VB_2B_3) \rightarrow X \neq B_3$. If $E = Q$, then (VEQ), $E \in A_1A_3$, $Q \in A_1A_2 \rightarrow E = Q = A$, whence $E \neq Q$. If $F = Q$, then (VQF), $Q \in B_2B_3$, $F \in B_1B_3 \rightarrow F = Q = B_3$, whence $F \neq Q$. Now, $A_1 \in FQ \rightarrow (A_1A_3V)$, so $A_1 \notin FQ$. But $X \in FQ$ or $B_3 \in FQ$ implies $A_1 \in FQ$, whence $X \notin FQ$ and $B_3 \notin FQ$. If (QA_1B_3), then $(QB_2B_3) \rightarrow (A_1B_2B_3)$, whence $\sim(QA_1B_3)$. To exclude (EA_1B_3) we first prove that $E \neq A_1$. If $E = A_1$, then (VQE) and $Q \in A_1A_2 \rightarrow Q = A$, whence $E \neq A_1$. Then (EA_1B_3), (EA_1A_3), $E \neq A_1 \rightarrow (A_1A_3B_3)$, whence $\sim(EA_1B_3)$. To exclude (FA_1B_3) we first prove that $F \neq B_3$. If $F = B_3$, then (VQF), (QB_2B_3), and $Q \neq B_3 \rightarrow (VB_2B_3)$, whence $F \neq B_3$. Then, (FA_1B_3), (FB_1B_3), and $F \neq B_3 \rightarrow (A_1B_1B_3)$, whence $\sim(FA_1B_3)$. This checks all the hypotheses of Definition 1 except $E \neq F$.

Subcase 1.1: $E \neq F$. Then, by Definition 1, $FX \wedge QA_1 = P_3$, $FB_3 \wedge A_1E = P_2$, and $QB_3 \wedge EX = P_1$ are collinear.

Subcase 1.2: (See Exercise 2.) $E = F$. Then, since $E \in A_1A_3$ and $F \in B_1B_3$, $E = F = P_2$. Recall that we already have (XEP_1) and (XFP_3), which now become (XP_2P_1), (XP_2P_3). Thus, if $X \neq P_2$, we have $(P_1P_2P_3)$. We show that $X = P_2$ is impossible. Now, $P_2 \neq A_1$, since $P_2 = A_1 \rightarrow (A_1B_1B_3)$. Then, if $X = P_2$, we would have (XA_1B_3), $(P_2A_1A_3)$, $X = P_2 \neq A_1 \rightarrow (A_1A_3B_3)$, whence $X \neq P_2$.

CASE 2. (See Exercise 3.) Case 1 does not hold. Then, by Lemma II, we may assume we have $(A_1B_2B_3)$, $(A_2B_1B_3)$, and $(A_3B_1B_2)$. As in Case 1, $P_i \neq A_j$, $P_i \neq B_j$ for all i, j. Let $Q_1 = B_3P_3 \wedge VA_1$, $Q_2 = B_3P_3 \wedge VA_2$. Since $P_3 \in B_1B_2$ and $P_3 \neq B_1, B_2, A_3$, we have $B_3P_3 \neq B_3A_1, B_3A_2, B_3A_3$. Therefore, points $P_3, Q_1, Q_2, V, A_1, A_2, A_3, B_1, B_2, B_3$ are distinct.

ASSERTION. Definition 1 applies to $V, A_3, B_3; P_3, A_2, A_1$. $\sim(B_1A_1A_2) \rightarrow V \notin P_3A_1$, $\sim(A_1A_2A_3) \rightarrow A_3 \notin P_3A$, $Q_1 \neq A_1 \rightarrow B_3 \notin P_3A_1$, $\sim(A_3B_3B_1) \rightarrow P_3 \notin VA_3$, $\sim(A_3B_3A_2) \rightarrow A_2 \notin VA_3$, and $\sim(VA_1A_3) \rightarrow A_1 \notin VA_3$.

Therefore, $(P_2B_2Q_1)$. The last assertion also allows us to apply Definition 1 to $V, A_3, B_3; P_3, A_1, A_2$ (A_1, A_2 are reversed!) and conclude $(P_1B_1Q_2)$.

ASSERTION. Definition 1 applies to Q_1, A_1, B_1; A_2, Q_2, B_2. For $\sim(A_1B_1A_2)$ implies Q_1, A_1, B_1 not on A_2Q_2 and A_2, Q_2, B_2 not on Q_1A_1.

Then, $A_1B_2 \wedge Q_2B_1 = P_1$, $A_2B_1 \wedge Q_1B_2 = P_2$, and $Q_1Q_2 \wedge A_1A_2 = P_3$ are collinear. ▌

THEOREM 2 $\mathbf{P}_2F$ is Pappian, where F is a field.

PROOF. Consider Figure III.5.1 in $\mathbf{P}_2F$. Let $AB \wedge A'B'\vdots\mathbf{p}$, $A\vdots\mathbf{a}_1$. Then $B\vdots\mathbf{p} + \gamma\mathbf{a}_1$ for some $\gamma \neq 0$. Let $\mathbf{a} = \gamma\mathbf{a}_1$. Then, $A\vdots\mathbf{a}$, $B\vdots\mathbf{p} + \mathbf{a}$, and then $C\vdots\mathbf{p} + \alpha\mathbf{a}$ for some $\alpha \neq 0, 1$. Similarly, there is $\mathbf{a}'$ and $\alpha' \neq 0, 1$ such that $A'\vdots\mathbf{a}'$, $B'\vdots\mathbf{p} + \mathbf{a}'$, $C'\vdots\mathbf{p} + \alpha'\mathbf{a}'$.

Let $\mathbf{q} = \mathbf{p} + \mathbf{a} + \mathbf{a}'$. $\mathbf{q} \neq \mathbf{0}$, $\mathbf{q}$ is a linear combination of $\mathbf{p} + \mathbf{a}$ and $\mathbf{a}'$, and $\mathbf{q}$ is a linear combination of $\mathbf{p} + \mathbf{a}'$ and $\mathbf{a}$. Therefore, $L = AB' \wedge A'B\vdots\mathbf{q}$. Similarly, $M = AC' \wedge A'C\vdots\mathbf{r} = \mathbf{p} + \alpha\mathbf{a} + \alpha'\mathbf{a}'$.

Let $\mathbf{s} = (1 - \alpha\alpha')\mathbf{p} + \alpha(1 - \alpha')\mathbf{a} + \alpha'(1 - \alpha)\mathbf{a}'$. $\mathbf{s} \neq \mathbf{0}$ because $\alpha \neq 0, 1$ and $\alpha' \neq 0, 1$. Also, $\mathbf{s} = (1 - \alpha')(\mathbf{p} + \alpha\mathbf{a}) + \alpha'(1 - \alpha)(\mathbf{p} + \mathbf{a}') = (1 - \alpha)(\mathbf{p} + \alpha'\mathbf{a}') + \alpha(1 - \alpha')(\mathbf{p} + \mathbf{a})$. Hence, $N = BC \wedge B'C'\vdots\mathbf{s}$.

Now $\alpha\alpha'\mathbf{q} - \mathbf{r} + \mathbf{s} = \mathbf{0}$. Therefore, L, M, N are collinear. ▌

The corresponding theorem for Desargues' configuration also holds for division rings (Exercise III.2.3). We see from the next result that this is not true for Pappus' configuration.

THEOREM 3 Every planar ternary ring of a Pappian projective plane has commutative multiplication.

PROOF. Let (S, T) be the planar ternary ring arising from the coordinatization of a given Pappian projective plane. Let $a, b \in S$, $a \neq b$, $b \neq 1$. In Figure III.5.1 the following points satisfy Definition 1. $A\vdots(1)$, $B\vdots(0, b)$, $C\vdots(b, 0)$; $A'\vdots(1, 0)$, $B'\vdots(a)$, $C'\vdots(0, a)$, $L\vdots(b)$, $M\vdots(a, 0)$, $N\vdots(0, ab)$. Then, since L, M, N are collinear, $N = LM \wedge BC'\vdots(b)(a, 0) \wedge [0] = (0, ba)$. Hence, $(0, ab) = (0, ba)$, and $ab = ba$. ▌

For the next result we require a special case of a theorem (IV.4.3) of analytic projective geometry.

LEMMA In $\mathbf{P}_2F$, F a field, let P_1, P_2, P_3, P_4 be four points, no three collinear, and let Q_1, Q_2, Q_3, Q_4 be four (possibly different) points, no three collinear. Then there exists a collineation ψ of $\mathbf{P}_2F$ such that $P_i\psi = Q_i$, $i = 1, 2, 3, 4$.

PROOF. Let $P_i\vdots\mathbf{p}_i'$, $i = 1, 2, 3, 4$. Since P_1, P_2, P_3 are not collinear, $\mathbf{p}_1', \mathbf{p}_2', \mathbf{p}_3'$ are independent. Then $\mathbf{p}_4' = \alpha_1\mathbf{p}_1' + \alpha_2\mathbf{p}_2' + \alpha_3\mathbf{p}_3'$ for some $\alpha_1, \alpha_2, \alpha_3$ in F. If $\alpha_1 = 0$, then P_2, P_3, P_4 are collinear, whence $\alpha_1 \neq 0$. Similarly, α_2 and α_3 are nonzero. Then $\alpha_i\mathbf{p}'_i$ is a coordinate vector for P_i, $i = 1, 2, 3$, so we may choose coordinate vectors so that

$$\mathbf{p}_1 + \mathbf{p}_2 + \mathbf{p}_3 = \mathbf{p}_4. \tag{1}$$

Similarly, we may choose $Q_i\!:\!\mathbf{q}_i$ so that

$$\mathbf{q}_1 + \mathbf{q}_2 + \mathbf{q}_3 = \mathbf{q}_4. \tag{2}$$

Let Ψ be the nonsingular 3×3 matrix satisfying $\Psi\mathbf{p}_i = \mathbf{q}_i$, $i = 1, 2, 3$. Then, by (1) and (2),

$$\Psi\mathbf{p}_i = \mathbf{q}_i, \qquad i = 1, 2, 3, 4. \tag{3}$$

Now define ψ as follows: If $P\!:\!\mathbf{p}$ is a point, let $P\psi\!:\!\Psi\mathbf{p}$; if $l\!:\!\mathbf{l}$ is a line, let $l\psi\!:\!\mathbf{l}\Psi^{-1}$. Since Ψ is nonsingular, ψ maps points and lines one–one onto points and lines. Since $\mathbf{l}\Psi^{-1}\Psi\mathbf{p} = \mathbf{l}\cdot\mathbf{p}$, ψ preserves incidence. Finally, by (3), $P_i\psi = Q_i$, $i = 1, 2, 3, 4$. ∎

This brings us to the last theorem of the sequence of configuration theorems.

THEOREM 4 A projective plane is Pappian if and only if it is isomorphic to $\mathbf{P}_2F$ for some field F. F is unique up to isomorphism when it exists.

PROOF. Suppose the projective plane $\mathbf{P}$ is isomorphic to $\mathbf{P}_2F$, where F is a field. By Theorem 2 $\mathbf{P}_2F$ is Pappian. Then, since the isomorphism and its inverse preserve incidence, $\mathbf{P}$ is Pappian.

Conversely, suppose $\mathbf{P}$ is Pappian. Let (F, T) be the planar ternary ring of some coordinatization of $\mathbf{P}$. By Theorem 1 $\mathbf{P}$ is Desarguesian. Then, by Theorem III.4.4, (F, T) is linear and $(F, +, \cdot)$ is a division ring. Then, by Theorem 3 $(F, +, \cdot)$ is a field. Define a function ϕ from the points and lines of $\mathbf{P}$ to the points and lines of $\mathbf{P}_2F$ as follows:*

$$\begin{aligned}
(x, y)\phi &= [(x, y, 1)],\\
(m)\phi &= [(1, -m, 0)],\\
(\infty)\phi &= [(0, 1, 0)],\\
[m, k]\phi &= [[m, 1, -k]],\\
[k]\phi &= [[1, 0, -k]],\\
l_\infty\phi &= [[0, 0, 1]],
\end{aligned}$$

where by $[\mathbf{v}]$ we mean $\{t\mathbf{v} \mid t \neq 0\}$. It is easy to check that ϕ is an isomorphism. For example: $(x, y) \in [m, k]$ iff $mx + y = k$ iff $(x, y, 1)\cdot[m, 1, -k] = 0$ iff $(x, y)\phi \in [m, k]\phi$. The reader should complete the check.

Now, to prove uniqueness, suppose the Pappian plane has isomorphisms θ and θ' onto $\mathbf{P}_2F$ and $\mathbf{P}_2F'$, respectively, where F and F' are fields. Let ϕ' be

---

* ϕ appears (essentially) at the end of Section III.1.

the map of **P** onto $\mathbf{P}_2F'$ corresponding to the map ϕ of **P** onto $\mathbf{P}_2F$ defined above. (Everything associated with F' will carry a prime.) By the lemma there is a collineation ψ of $\mathbf{P}_2F$ sending (0, 0, 1), (1, 0, 1), (1, −1, 0), (0, 1, 0)

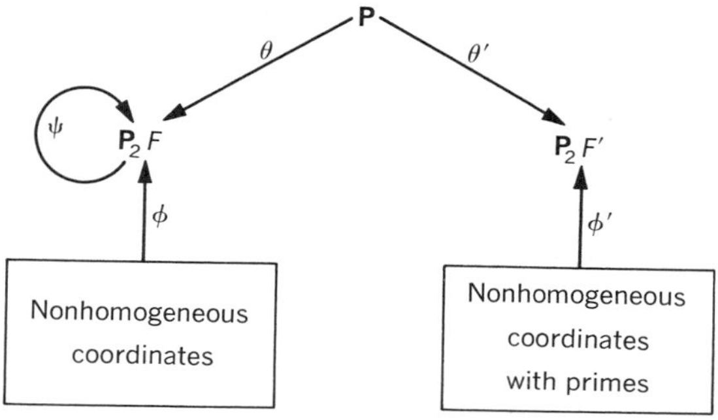

Fig. III.5.3

to $(0', 0', 1')\theta'^{-1}\theta$, $(1', 0', 1')\theta'^{-1}\theta$, $(1', -1', 0')\theta'^{-1}\theta$, $(0', 1', 0')\theta'^{-1}\theta$, respectively (see Figure III.5.3). Let $\mu = \phi\psi\theta^{-1}\theta'\phi'^{-1}$. Then

$$(0, 0)\mu = (0', 0'),$$
$$(1, 0)\mu = (1', 0'),$$
$$(1)\mu = (1'),$$
$$(\infty)\mu = (\infty'),$$

and, hence,

$$(0)\mu = (0'),$$
$$[0]\mu = [0'],$$
$$[0, 0]\mu = [0', 0'],$$
$$l_\infty\mu = l'_\infty.$$

Now define a function α: $F \to F'$ as follows: If $x \in F$, then $(x, 0) \in [0, 0]$ and $(x, 0) \neq (0)$. Hence, $(x, 0)\mu \in' [0', 0']$ and $(x, 0)\mu \neq (0')$. Therefore, $(x, 0)\mu = (x', 0')$ for some unique x' in F'. Let $x' = x\alpha$. Since μ is one–one and onto, so is α. Given $(a, 0)$ and $(b, 0)$, the points $(ab, 0)$ and $(a + b, 0)$ are determined solely by the points (0, 0), (1, 0), (1), and (∞), and by the incidence properties of **P** (Figure III.5.4). Since these things are all preserved by μ, the configurations in Figure III.5.4 still hold when decorated with primes. Thus, $(ab)' = a'b'$ and $(a + b)' = a' + b'$, that is, α is an isomorphism. ∎

A reader who is familiar with linear algebra over division rings will have little trouble extending this result to Desarguesian planes and $\mathbf{P}_2D$, D a division ring (see end of Section 4).

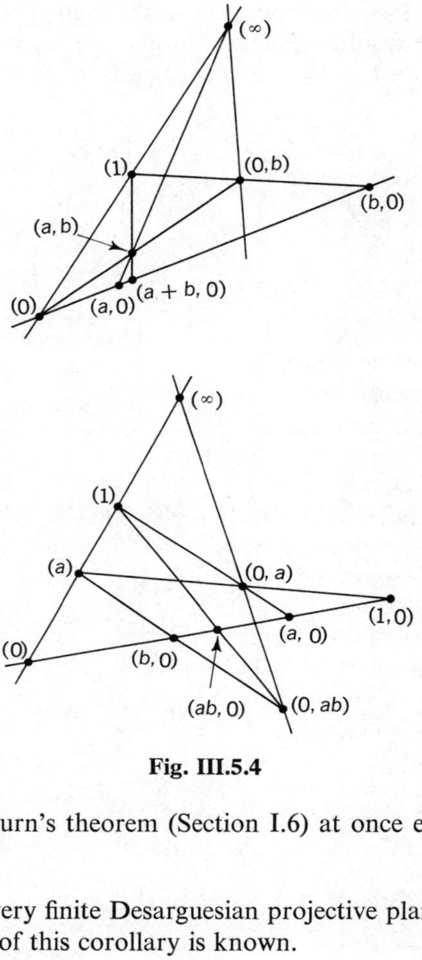

Fig. III.5.4

Finally, Wedderburn's theorem (Section I.6) at once extracts a corollary from Theorem 4.

COROLLARY Every finite Desarguesian projective plane is Pappian.
No synthetic proof of this corollary is known.

Exercises

1. Show that in any meaningful setup of Figure III.5.1, not covered by Definition 1 (for example, $A = B$ or C on $A'B'$), the conclusion L, M, N collinear holds in any projective plane.
2. Show by explicit example in $\mathbf{P}_2\mathrm{R}$ that Subcase 1.2 in the proof of Theorem 1 can actually occur.
3. Show by explicit example in $\mathbf{P}_2\mathrm{R}$ that Case 2 in the proof of Theorem 1 can actually occur.

4. Using the Artin–Zorn theorem (Section I.6), show that every finite Moufang plane is Desarguesian.

6. HIGHER-DIMENSIONAL SPACES

Our investigations thus far have been restricted to projective spaces of dimension two, but not because spaces of higher dimension are more difficult to manage. On the contrary, as we shall see, spaces of dimension other than two are *much easier* to study than planes.

The reader may need to refer to Section II.3 for terms and results employed below.

Projective spaces of dimensions -1, 0, or 1 have no interest in themselves, but are studied only as subspaces of higher-dimensional spaces. For this reason the term "space" will be taken throughout this section to mean "projective space of dimension greater than two."

Spaces have relatively simple structures because the classical proof of Desargues' theorem can be used in them. We present this as a lemma.

LEMMA Let P, A, B, C, A', B', C' be distinct points in a three-dimensional space such that

(1) P, A, A' are collinear, P, B, B' are collinear, and P, C, C' are collinear;
(2) A', B', C' are not collinear;
(3) P, A, B, C are not coplanar.

Then $AB \wedge A'B'$, $AC \wedge A'C'$, $BC \wedge B'C'$ exist and are collinear.

PROOF. Let $\mathbf{P} = \mathbf{P}(C, AB)$, $\mathbf{P}' = \mathbf{P}(C', A'B')$. If $\mathbf{P} = \mathbf{P}'$, we would have A and A' in $\mathbf{P}$ and by (1) P, A, B, C in $\mathbf{P}$, contradicting (3). Therefore, $\mathbf{P} \neq \mathbf{P}'$. Now, A, B, A', B' are in $\mathbf{P}(P, AB)$, so $D = AB \wedge A'B'$ exists. Similarly, $E = AC \wedge A'C'$ and $F = BC \wedge B'C'$ exist. By the corollary to Theorem II.3.8, $\mathbf{P} \wedge \mathbf{P}'$ is a line l. Since $D \in AB$ and $D \in A'B'$, D is on $\mathbf{P} \wedge \mathbf{P}'$, that is, $D \in l$. Similarly, $E \in l$ and $F \in l$. ▎

THEOREM 1 Every plane in a space is Desarguesian.

PROOF. Let P, A, B, C, A', B', C' be the usual setup for Desargues' configuration in a plane $\mathbf{P}$ in some space (Figure III.6.1). Let P^* be a point not on $\mathbf{P}$. We proceed in the three-dimensional space $\mathbf{P}_3(P^*, \mathbf{P})$. Let A^* be a point on P^*A distinct from P^* and A. (PA^* and P^*A' are distinct lines in $\mathbf{P}(P^*, PA')$. Let $A^{*\prime} = PA^* \wedge P^*A'$. Applying the lemma to $P, A^*, B, C, A^{*\prime}, B', C'$, we have that $L^* = A^*B \wedge A^{*\prime}B'$, $M^* = A^*C \wedge A^{*\prime}C'$, and $N = BC \wedge B'C'$ exist and are collinear. Let $\mathbf{P}^* = \mathbf{P}(P^*, L^*M^*)$.

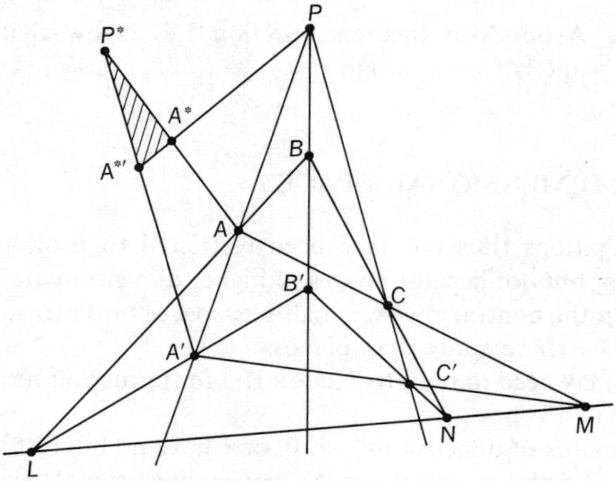

Fig. III.6.1

L, L^* and P^* are on both $\mathbf{P}(P^*, AB)$ and $\mathbf{P}(P^*, A'B')$, hence, $L \in P^*L^*$; therefore, L is on $\mathbf{P}^*$.

M, M^* and P^* are on both $\mathbf{P}(P^*, AC)$ and $\mathbf{P}(P^*, A'C')$, hence, $M \in P^*M^*$; therefore, M is on $\mathbf{P}^*$.

$N \in L^*M^*$, therefore, N is on $\mathbf{P}^*$.

We have shown that L, M, N are on $\mathbf{P}^*$. Thus, L, M, N are on the line $\mathbf{P} \wedge \mathbf{P}^*$. ▎

In the next theorem we prove in detail a special case of the general results suggested by Theorem 1.

THEOREM 2 If one plane of a projective 3-space $\mathbf{P}_3$ is Pappian, then $\mathbf{P}_3$ is isomorphic to $\mathbf{P}_3F$ for a unique field F.

PROOF. Let $\mathbf{P}$ be a Pappian plane in $\mathbf{P}_3$. By Theorem 5.4, $\mathbf{P}$ is isomorphic to $\mathbf{P}_2F$ for a unique field F. If $\mathbf{P}'$ is any other plane in $\mathbf{P}_3$, then, since $\mathbf{P}$ is isomorphic to $\mathbf{P}'$ (Exercise II.5.6), $\mathbf{P}'$ is isomorphic to $\mathbf{P}_2F$.

Let $\mathbf{P}_{xy}$, $\mathbf{P}_{xz}$, $\mathbf{P}_{yz}$ be three planes having no line in common. Let $a_x = \mathbf{P}_{xy} \wedge \mathbf{P}_{xz}$, $a_y = \mathbf{P}_{xy} \wedge \mathbf{P}_{yz}$, $a_z = \mathbf{P}_{xz} \wedge \mathbf{P}_{yz}$, $O = a_x \wedge a_y \wedge a_z$. Choose points $A_x \in a_x$, $A_y \in a_y$, $A_z \in a_z$ distinct from O. Let $b_{\infty x} = A_yA_z$, $b_{\infty y} = A_xA_z$, $b_{\infty z} = A_xA_y$. In $\mathbf{P}_\infty = \mathbf{P}(A_z, b_{\infty z})$ choose a line k not through A_x, A_y, or A_z. Let $I_x = k \wedge b_{\infty x}$, $I_y = k \wedge b_{\infty y}$, $I_z = k \wedge b_{\infty z}$.

Coordinatize $\mathbf{P}_{xy}$ with F so that

$$a_x \vdots [0, 0],$$

$$a_y \vdots [0],$$

$$b_{\infty z} \vdots l_\infty,$$

$$I_z \vdots (1).$$

(Except for these restrictions, this can be done in any way.) Coordinatize $\mathbf{P}_{yz}$ so that

$$a_y : [0, 0],$$
$$a_z : [0],$$
$$b_{\infty x} : l_\infty,$$
$$I_x : (1),$$

and so that a point with $\mathbf{P}_{xy}$-coordinates $(0, y)$ receives $\mathbf{P}_{yz}$-coordinates $(y, 0)$. Coordinatize $\mathbf{P}_{xz}$ so that

$$a_x : [0, 0],$$
$$a_z : [0],$$
$$b_{\infty y} : l_\infty,$$
$$I_y : (1),$$

and so that a point with $\mathbf{P}_{xy}$-coordinates $(x, 0)$ receives $\mathbf{P}_{xz}$-coordinates $(x, 0)$. Then, since I_x, I_y, I_z are collinear (on k), a point with $\mathbf{P}_{yz}$-coordinates $(0, z)$ will receive $\mathbf{P}_{xz}$-coordinates $(0, z)$ too. Now, for any point P not on $\mathbf{P}_\infty$ suppose $\mathbf{P}(P, b_{\infty x}) \wedge a_x : (x, 0)$ in $\mathbf{P}_{xy}$, $\mathbf{P}(P, b_{\infty y}) \wedge a_y : (y, 0)$ in $\mathbf{P}_{yz}$, $\mathbf{P}(P, b_{\infty z}) \wedge a_z : (0, z)$ in $\mathbf{P}_{xz}$. Then let $P : (x, y, z)$. This assigns to all points of $\mathbf{P}_{xy}$ not on $b_{\infty z}$ coordinates $(x, y, 0)$, to points of $\mathbf{P}_{yz}$ not on $b_{\infty x}$ coordinates $(0, y, z)$, and to points of $\mathbf{P}_{xz}$ not on $b_{\infty y}$ coordinates $(x, 0, z)$, with proper "matching up" on a_x, a_y, and a_z.

Now replace each coordinate triple (x, y, z) by the usual homogeneous coordinates $\{t(x, y, z, 1) \mid t \neq 0\}$. Let l be a line not on $\mathbf{P}_\infty$. Let A, B, P be points on l not on $\mathbf{P}_\infty$, say $A : \mathbf{a} = (a_1, a_2, a_3, 1)$, $B : \mathbf{b} = (b_1, b_2, b_3, 1)$, $P : \mathbf{p} = (x_1, x_2, x_3, 1)$. To show that the portion of $\mathbf{P}_3$ not on $\mathbf{P}_\infty$ coincides with $\mathbf{P}_3F$ minus the plane $x_4 = 0$, we must show that each line in $\mathbf{P}_3$ not in $\mathbf{P}_\infty$ coincides with a linear subspace of dimension 1 in $\mathbf{P}_3F$. With the above setup, this means we must exhibit the coordinate vector $\mathbf{p}$ as a linear combination of $\mathbf{a}$ and $\mathbf{b}$.

Now l cannot pass through two of the points A_x, A_y, A_z, so we may assume it does not pass through A_x or A_z. Let $l' = \mathbf{P}(A_z, l) \wedge \mathbf{P}_{xy}$. Then $A' = A_zA \wedge \mathbf{P}_{xy}$, $B' = A_zB \wedge \mathbf{P}_{xy}$, $P' = A_zP \wedge \mathbf{P}_{xy}$ are points on l'. By our coordinatization then $A' : (a_1, a_2, 1)$, $B' : (b_1, b_2, 1)$, $P : (x_1, x_2, 1)$ in $\mathbf{P}_{xy}$. (Actually, points in $\mathbf{P}_{xy}$ not on $b_{\infty z}$ have coordinates of the form $(x, y, 0, 1)$. But the third-place entry 0 has no effect in $\mathbf{P}_{xy}$, so we drop it.) Then, since $\mathbf{P}_{xy}$ is a $\mathbf{P}_2F$, there are p, q, F such that

$$pa_1 + qb_1 = x_1,$$
$$pa_2 + qb_2 = x_2,$$
$$p + q = 1.$$

Similarly, considering $l'' = \mathbf{P}(A_x, l) \wedge \mathbf{P}_{yz}$, there is p', q' in F such that

$$p'a_2 + q'b_2 = x_2,$$
$$p'a_3 + q'b_3 = x_3,$$
$$p' + q' = 1.$$

Comparing these equations with those above, we see that $p = p'$ and $q = q'$; hence,

$$\mathbf{p} = p\mathbf{a} + q\mathbf{b}.$$

Finally, since $\mathbf{P}_3$ and $\mathbf{P}_3F$ agree everywhere off $\mathbf{P}_\infty$, they are the same (Exercise 1). ▎

The argument just concluded can be modified and extended to prove the full results stated in the next theorem. However, we shall not do this.

THEOREM 3 Every projective n-space $\mathbf{P}_n$, where n is greater than two, is isomorphic to $\mathbf{P}_nD$ for a unique division ring D. If some plane of $\mathbf{P}_n$ is Pappian, then D is a field.

The study of higher-dimensional spaces is, then, just an application of linear algebra over a division ring. In the next chapter we shall present the most important basic results of this study, under the restriction that the division ring be a field—a restriction which we make so as not to exceed the bounds of the usual first course in linear algebra. For the study in complete generality see [3].

Exercises

1. Let $\mathbf{P}_\infty$ be a plane in a projective 3-space $\mathbf{P}_3$ and $\mathbf{P}'_\infty$ a plane in a projective 3-space $\mathbf{P}'_3$. Suppose there is an isomorphism from the points and lines of $\mathbf{P}_3$ not on $\mathbf{P}_\infty$ onto the points and lines of $\mathbf{P}'_3$ not on $\mathbf{P}'_\infty$. Show that this isomorphism may be extended to an isomorphism of $\mathbf{P}_3$ onto $\mathbf{P}'_3$.
2. (Toward a proof of Theorem 3.) Prove that any two planes in a projective space are isomorphic.
3. Prove that any three noncollinear points in a projective 3-space can be sent to any other three noncollinear points by a product of projections. How many projections are in general needed? (For the 2-space case, see Exercise II.1.3.)

*4. Let F be a finite field with exactly q elements. Prove that the number of r-dimensional linear subspaces of $\mathbf{P}_nF$ ($r \le n$) is

$$\frac{(q^{n+1} - 1)(q^n - 1) \cdots (q^{n-r+1} - 1)}{(q^{r+1} - 1)(q^r - 1) \cdots (q - 1)}.$$

IV
Analytic Projective Spaces

The reader should now read Section I.6 and restudy Sections I.3 and I.4. Throughout this chapter F denotes an arbitrary field.

1. HYPERPLANE COORDINATES AND DUALITY

Let $\mathbf{P}_{n-1}$ be a hyperplane in $\mathbf{P}_nF$ (that is, an $(n-1)$-dimensional subspace). By Theorem II.4.5, there is an $(n+1) \times (n+1)$ matrix A over F of rank one such that a point P is in $\mathbf{P}_{n-1}$ if and only if $\mathbf{p}A = \mathbf{0}$ for some (and, hence, every) coordinate vector $\mathbf{p}$ of P. Choose a nonzero column vector $\mathbf{u}$ of A. Then, since A is of rank one, $\mathbf{p}A = \mathbf{0}$ if and only if $\mathbf{pu} = 0$. Thus $\mathbf{P}_{n-1}$ consists of all points $P{:}\mathbf{p}$ such that

$$\mathbf{pu} = 0 \tag{1}$$

DEFINITION 1 The homogeneous *hyperplane coordinates* of a hyperplane with Eq. (1) are $\{t\mathbf{u} \mid t \neq 0\}$.

In $\mathbf{P}_2F$ hyperplanes are just lines, and hyperplane coordinates are the line coordinates of Section II.2. (Hyperplane coordinates are usually written $[u_0, u_1, \cdots, u_n]$, as were line coordinates in Chapter II.) Clearly, an n-dimensional duality theorem is in order.

THEOREM 1 (Duality in n-space.) Given any theorem $\mathfrak{T}$ about linear subspaces in $\mathbf{P}_nF$, the *dual* statement $\mathfrak{T}^*$ obtained by interchanging "dimension r" with "dimension $n - r - 1$" and "contains" with "is contained in" is also a theorem.

Before proving this theorem we require some discussion.

DEFINITIONS

2. The *linear bundle* of hyperplanes *spanned* by hyperplanes $\mathbf{P}^1, \mathbf{P}^2, \cdots, \mathbf{P}^m$ is the set of all hyperplanes having coordinate vectors that are linear combinations of coordinate vectors $\mathbf{u}_1, \mathbf{u}_2, \cdots, \mathbf{u}_m$ of $\mathbf{P}^1, \mathbf{P}^2, \cdots, \mathbf{P}^m$, respectively.
3. The *dimension* of the bundle spanned by $\mathbf{P}^1, \cdots, \mathbf{P}^m$ is the dimension of the vector space spanned by $\mathbf{u}_1, \cdots, \mathbf{u}_m$ minus one.

4. The *center* of the bundle is the intersection of all the hyperplanes in the bundle.

THEOREM 2 The center of an r-dimensional bundle of hyperplanes in $\mathbf{P}_nF$ is an $(n - r - 1)$-dimensional subspace of $\mathbf{P}_nF$.

PROOF. Let B be an r-dimensional bundle. Just as in Theorem II.4.5, there is a matrix A of rank $n - r$ such that B consists of all hyperplanes having coordinate vectors $\mathbf{u}$ such that $A\mathbf{u} = \mathbf{0}$. Pick $n - r$ independent row vectors $\mathbf{p}_1, \cdots, \mathbf{p}_{n-r}$ of A. Let $\mathbf{C}$ be the $(n - r - 1)$-dimensional linear subspace spanned by points $P_1{:}\mathbf{p}_1, \cdots, P_{n-r}{:}\mathbf{p}_{n-r}$. Then a point P is in $\mathbf{C}$ iff P has a linear combination $\mathbf{p}$ of the $\mathbf{p}_i$ as coordinate vector, iff $\mathbf{pu} = 0$ for every $\mathbf{u}$ such that $A\mathbf{u} = \mathbf{0}$, iff P is on every hyperplane of B, iff P is in the center of B. Thus $\mathbf{C}$ is the center of B. ▎

The proof of Theorem 2 immediately yields a corollary.

COROLLARY Let B_1, B_2 be bundles with centers $\mathbf{C}_1$, $\mathbf{C}_2$ respectively. Then $B_1 \subset B_2$ if and only if $\mathbf{C}_1 \supset \mathbf{C}_2$.

PROOF OF THEOREM 1. Let $\mathfrak{T}$ be a theorem about linear subspaces in $\mathbf{P}_nF$. We replace each point coordinate vector $\mathbf{p}$ by its transpose $\mathbf{p}^T$, and regard the transposed coordinate sets $\{t\mathbf{p}^T \mid t \neq 0\}$ as coordinates for hyperplanes. $\mathfrak{T}$ then becomes a theorem about linear bundles of hyperplanes. Then the statement implied by $\mathfrak{T}$ and Theorem 2 and its corollary is a case of $\mathfrak{T}^*$. Finally, that every case of $\mathfrak{T}^*$ can be obtained in this way follows from Exercise 1. ▎

In applying this theorem, we will usually just say "by duality $\cdots$" or "dually $\cdots$."

Exercises

1. Prove the converse of Theorem 2: The set of all hyperplanes containing an r-dimensional subspace of $\mathbf{P}_nF$ forms a linear bundle of dimension $n - r - 1$.
2. Generalize Exercise II.2.4.
3. What is the minimum dimension of the intersection of r hyperplanes in $\mathbf{P}_nF$?
4. Generalize Exercise II.2.9.
5. Show how the $(r + 1)$-dimensional subspaces of $\mathbf{P}_nF$ containing a fixed r-dimensional subspace can be considered as the points of an $(n - r - 1)$-dimensional subspace of $\mathbf{P}_nF$.

2. FRAME OF REFERENCE AND CROSS RATIO

In the next chapter we will often find it convenient to re-coordinatize a subspace of a $\mathbf{P}_nF$ in order to simplify certain calculations. In this section we present a good way to accomplish this, and as a by-product we obtain one of the basic tools of classical analytic projective geometry.

* * *

DEFINITION 1 Let $\mathbf{P}_r$ be an r-dimensional subspace of $\mathbf{P}_nF$. A *simplex* in $\mathbf{P}_r$ is an ordered set of $r + 1$ independent points in $\mathbf{P}_r$. A *proper face* of a simplex $(Q_0, Q_1, \cdots, Q_r)$ in P_r is any subspace of $\mathbf{P}_r$ that is spanned by a proper subset of $\{Q_0, Q_1, \cdots, Q_r\}$.

THEOREM 1 Let $(Q_0, Q_1, \cdots, Q_r)$ be a simplex in an r-dimensional subspace $\mathbf{P}_r$ of $\mathbf{P}_nF$. Let U be a point of $\mathbf{P}_r$ that is in no proper face of the simplex. Then there exist coordinate vectors $U\colon\mathbf{u}$, $Q_i\colon\mathbf{q}_i$, $0 \leq i \leq r$, such that

$$\mathbf{u} = \mathbf{q}_0 + \mathbf{q}_1 + \cdots + \mathbf{q}_r. \tag{1}$$

Moreover, if $U\colon\mathbf{u}'$, $Q_i\colon\mathbf{q}_i'$ is another choice of coordinate vectors satisfying (1), then there exists a $t \neq 0$ in F such that

$$\mathbf{u}' = t\mathbf{u}, \qquad \mathbf{q}_i' = t\mathbf{q}_i, \qquad 0 \leq i \leq r. \tag{2}$$

PROOF. Let $U\colon\mathbf{u}$, $Q_i\colon\mathbf{q}_i^*$ be any choice of coordinate vectors. Since $Q_0, Q_1, \cdots, Q_r$ span $\mathbf{P}_r$, there exist $\alpha_0, \alpha_1, \cdots, \alpha_r$ in F such that $\mathbf{u} = \alpha_0\mathbf{q}_0^* + \alpha_1\mathbf{q}_1^* + \cdots + \alpha_r\mathbf{q}_r^*$. Since U is in no proper face, $\alpha_i \neq 0$ for all i. Thus, $\mathbf{q}_i = \alpha_i\mathbf{q}_i^*$ is a coordinate vector for Q_i, and (1) is satisfied.

Suppose $\mathbf{u}'$, $\mathbf{q}_i'$ is another choice of coordinate vectors satisfying (1). Now $\mathbf{u}' = t\mathbf{u}$, $\mathbf{q}_i' = t_i\mathbf{q}_i$ for some $t_i \neq 0$ in F. Then $\sum \mathbf{q}_i = \mathbf{u} = t^{-1}\mathbf{u}' = \sum t^{-1}\mathbf{q}_i' = \sum (t^{-1}t_i)\mathbf{q}_i$ and, since the $\mathbf{q}_i$ are independent, $t^{-1}t_i = 1$ for all i. Hence, $\mathbf{q}_i' = t\mathbf{q}_i$ for all i. ∎

A simplex $(Q_0, Q_1, \cdots, Q_r)$ in $\mathbf{P}_r$ can be used to re-coordinatize $\mathbf{P}_r$ as follows: Choose a point U that is in no proper face of the simplex. Let $\mathbf{u}$, $\mathbf{q}_i$ be a choice of coordinate vectors satisfying (1). Let P be any point in P_r and let $\mathbf{p}$ be any coordinate vector for P. Then $\mathbf{p} = \sum x_i\mathbf{q}_i$ for a unique $\mathbf{x} = (x_0, x_1, \cdots, x_r)$ in F^{r+1}. If a different $\mathbf{p}$ is selected, $\mathbf{x}$ is still unique up to a nonzero multiple. If other $\mathbf{u}$, $\mathbf{q}_i$ are selected [still satisfying (1)], then, by (2), $\mathbf{x}$ is *still* unique up to a nonzero multiple. Thus, for any P the set

$$\mathcal{C}_p = \{(x_0, x_1, \cdots, x_r) \mid P\colon \textstyle\sum x_i\mathbf{q}_i \text{ and } U\colon \sum \mathbf{q}_i\} \tag{3}$$

is uniquely determined by P, the simplex, and the point U. Moreover, if $\mathbf{x}$ is any element of $\mathcal{C}_p$, then $\mathcal{C}_p = [\mathbf{x}] = \{t\mathbf{x} \mid t \neq 0\}$.

DEFINITION 2 The set (3) is the coordinate set of P in the *frame of reference* $(Q_0Q_1 \cdots Q_r \mid U)$ with *coordinate simplex* $(Q_0, Q_1, \cdots, Q_r)$ and *unit point* U.

Note that in the frame $(Q_0Q_1 \cdots Q_r \mid U)$ Q_i has the coordinate $(0, 0, \cdots, 1, \cdots, 0)$, with the 1 in the $(i + 1)$st place, and U has the coordinate $(1, 1, \cdots, 1)$. Let us now examine the effect of changing from one frame to another.

THEOREM 2 Suppose a point has the coordinate set $[\mathbf{x}]$ in one frame of reference and $[\mathbf{x}']$ in another one. Then there exists a matrix A such that $[\mathbf{x}'] = [A\mathbf{x}]$. A is nonsingular and is unique up to a constant multiple.

PROOF. Let $P\colon\mathbf{p}$ be the point, $(Q_0Q_1 \cdots Q_r \mid U)$ the first frame, $(Q'_0Q'_1 \cdots Q'_r \mid U')$ the second. Choose coordinate vectors $Q_i\colon\mathbf{q}_i$, $U\colon\mathbf{u}$ such that $\mathbf{u} = \sum \mathbf{q}_i$ and $\mathbf{p} = \sum x_i\mathbf{q}_i$, where $\mathbf{x} = (x_0, x_1, \cdots, x_r)$. Choose coordinate vectors $Q'_i\colon\mathbf{q}'_i$, $U'\colon\mathbf{u}'$ such that $\sum \mathbf{q}'_i = \mathbf{u}'$. Let $\mathbf{x}' = (x'_0, x'_1, \cdots, x'_r)$. $\sum x'_i\mathbf{q}'_i$ is a coordinate vector for P, so $\sum x'_i\mathbf{q}'_i = t \sum x_i\mathbf{q}_i$ for some $t \neq 0$ in the field. Since $\{\mathbf{q}'_i\}$ is a basis for the $(r + 1)$-dimensional vector space containing $\{\mathbf{q}_i\}$ there are unique constants a_{ij}, $0 \leq i \leq r$, $0 \leq j \leq r$, such that $\mathbf{q}_i = \sum_j a_{ji}\mathbf{q}'_j$, $0 \leq i \leq r$. Since the $\mathbf{q}_i$ are independent, the matrix $A = (a_{ij})$ is nonsingular. Then $t^{-1} \sum_i x'_i\mathbf{q}'_i = \sum_i x_i\mathbf{q}_i = \sum_i x_i(\sum_j a_{ji}\mathbf{q}'_j) = \sum_j (\sum_i x_ia_{ji})\mathbf{q}'_j = \sum_i (\sum_j x_ja_{ij})\mathbf{q}'_i$, where in the last step we have merely interchanged the letters i and j. By independence of $\{\mathbf{q}'_i\}$ then $t^{-1}x'_i = \sum_j a_{ij}x_j$, $0 \leq i \leq r$, that is,

$$\mathbf{x}' = tA\mathbf{x}.$$

Hence, $[\mathbf{x}'] = [A\mathbf{x}]$. Obviously, any nonzero scalar multiple of A will work as well as A. By Theorem 1 only such scalar multiples will do. ∎

In Section II.6, we discussed briefly the affine coordinate system available in $\mathbf{P}_nF$ with the hyperplane $[0, 0, \cdots, 1]$ removed. We are now ready to extend and complete this discussion.

Let $\mathbf{P}_r$ be an r-dimensional subspace of $\mathbf{P}_nF$. Let $\mathbf{P}_{r-1}$ be a hyperplane in $\mathbf{P}_r$, that is, an $(r - 1)$-dimensional subspace of $\mathbf{P}_nF$ that is contained in $\mathbf{P}_r$. Let $\{Q_1, Q_2, \cdots, Q_r\}$ be a basis for $\mathbf{P}_{r-1}$. Let Q_0 be any point in $\mathbf{P}_r$ not on $\mathbf{P}_{r-1}$. Then $(Q_0, Q_1, \cdots, Q_r)$ is a simplex in $\mathbf{P}_r$. Choose a unit point U. As explained above, a point P in $\mathbf{P}_r$ receives coordinates $\{t(x_0, x_1, \cdots, x_r) \mid t \neq 0\}$ in the frame $(Q_0Q_1 \cdots Q_r \mid U)$. P is on $\mathbf{P}_{r-1}$ if and only if $x_0 = 0$. Thus for every P not on $\mathbf{P}_{r-1}$ the r-tuple

$$x_0^{-1}(x_1, x_2, \cdots, x_r) \tag{4}$$

is uniquely determined by $(Q_0Q_1 \cdots Q_r \mid U)$; r-tuple (4) is the *affine coordinate* of P relative to the frame. This will be useful in the study of curves and surfaces (Chapter V). It also yields a particularly useful concept in the one-dimensional case.

DEFINITION 3 Let P_1, P_2, P_3 be distinct collinear points in a $\mathbf{P}_nF$ and let P_4 be a point on the same line that is distinct from P_1. The affine coordinate of P_4 in the frame $(P_1P_2|P_3)$ is called the *cross ratio* $(P_1P_2P_3P_4)$.

Notice that the only requirement for three points to form a frame of reference is that they be distinct and collinear. Also note that the affine space of a frame $(P_1P_2|P_3)$ on a line l consists of all points on l except P_1. Thus, the qualifications in the definition are both necessary and sufficient. What are the possible values of a cross ratio?

THEOREM 3 Let P_1, P_2, P_3 be distinct points on a line l in $\mathbf{P}_nF$. For each x in F there is a unique P on l. $P \neq P_1$, such that

$$(P_1P_2P_3P) = x. \tag{5}$$

PROOF. Pick coordinate vectors such that $\mathbf{p}_1 + \mathbf{p}_2 = \mathbf{p}_3$. Then $(P_1P_2P_3P) = x$ if and only if P is the point having coordinate vector $x\mathbf{p}_1 + \mathbf{p}_2$. ∎
The particular solutions of (5) for $x = 0$, 1 are $P = P_2$, P_3, respectively. Thus, the following corollary holds.

COROLLARY P_1, P_2, P_3, P_4 are distinct points on a line if and only if $(P_1P_2P_3P_4)$ is defined and different from 0 and 1.

We next obtain a useful expression for the cross ratio in terms of affine coordinates.

THEOREM 4 Let P_1, P_2, P_3 be distinct points on a line l. Let $P_4 \neq P_1$ be a point on l. Let $(Q_0Q_1|U)$ be a frame of reference on l, with $Q_0 \neq P_i$, $1 \leq i \leq 4$. Suppose P_i has affine coordinate x_i in this frame, $1 \leq i \leq 4$. Then

$$(P_1P_2P_3P_4) = \frac{(x_1 - x_3)(x_2 - x_4)}{(x_1 - x_4)(x_2 - x_3)}. \tag{6}$$

PROOF. By hypothesis there are coordinate vectors $P_i{:}\mathbf{p}_i$, $1 \leq i \leq 4$, $Q_j{:}\mathbf{q}_j$, $j = 1, 2$, $U{:}\mathbf{u}$ such that $\mathbf{q}_0 + \mathbf{q}_1 = \mathbf{u}$ and $\mathbf{p}_i = x_i\mathbf{q}_0 + \mathbf{q}_1$, $1 \leq i \leq 4$. Let $\mathbf{p}_1' = (x_3 - x_2)\mathbf{p}_1$, $\mathbf{p}_2' = (x_1 - x_3)\mathbf{p}_2$, $\mathbf{p}_3' = (x_1 - x_2)\mathbf{p}_3$, $\mathbf{p}_4' = (x_1 - x_2)\mathbf{p}_4$. The reader should check by substitution, multiplication and cancellation that $\mathbf{p}_1' + \mathbf{p}_2' = \mathbf{p}_3'$ and $\mathbf{p}_4' = (x_4 - x_2)(x_3 - x_2)^{-1}\mathbf{p}_1' + (x_1 - x_4)(x_1 - x_3)^{-1}\mathbf{p}_2'$. Then the affine coordinate of P_4 in $(P_1P_2|P_3)$ is $(P_1P_2P_3P_4) = (x_4 - x_2)(x_3 - x_2)^{-1}/(x_1 - x_4)(x_1 - x_3)^{-1} = (x_1 - x_3)(x_2 - x_4)/(x_1 - x_4)(x_2 - x_3)$. ∎

COROLLARY Let P_1, P_2, P_3, P_4 be distinct collinear points. Let $k = (P_1P_2P_3P_4)$. Then $k \neq 0, 1$ and

$$(P_1P_2P_3P_4) = (P_2P_1P_4P_3) = (P_3P_4P_1P_2) = (P_4P_3P_2P_1) = k,$$
$$(P_1P_2P_4P_3) = (P_2P_1P_3P_4) = (P_4P_3P_1P_2) = (P_3P_4P_2P_1) = k^{-1},$$
$$(P_1P_3P_2P_4) = (P_3P_1P_4P_2) = (P_2P_4P_1P_3) = (P_4P_2P_3P_1) = 1 - k,$$
$$(P_1P_3P_4P_2) = (P_3P_1P_2P_4) = (P_4P_2P_1P_3) = (P_2P_4P_3P_1) = (1 - k)^{-1},$$
$$(P_1P_4P_2P_3) = (P_4P_1P_3P_2) = (P_2P_3P_1P_4) = (P_3P_2P_4P_1) = (k - 1)k^{-1},$$
$$(P_1P_4P_3P_2) = (P_4P_1P_2P_3) = (P_3P_2P_1P_4) = (P_2P_3P_4P_1) = k(k - 1)^{-1}.$$

PROOF. Exercise 1.

Exercises*

1. Prove the corollary to Theorem 4 by using (6). (The tidiest proof makes use of the group S_4 of all permutations of $\{P_1, P_2, P_3, P_4\}$.)
2. In general the six values of the cross ratio of four distinct points are distinct numbers. Prove that if the characteristic of F is not three, then only three of these values are distinct if and only if $k = -1$, $k = 2$, $k = \frac{1}{2}$, or k is one of the two distinct roots (if they exist in F) of $x^2 - x + 1$. Prove that if F is of characteristic three, then $x^2 - x + 1$ has the repeated root 2, and either the six values are distinct or they all equal 2. In a field of characteristic two the cases $k = -1, 2, \frac{1}{2}$ are excluded, but $x^2 - x + 1$ may or may not have roots.
3. As x ranges over F the point P of (5) ranges over $\tilde{l} - \{P_1\}$. The value ∞ is often assigned to $(P_1P_2P_3P_1)$. Show, using (6) and the usual properties of the symbol ∞, that the corollary to Theorem 4 has meaning in this case. Can other cases, such as $P_2 = P_3$, be given meaning?

3. COMPLETE QUADRANGLES

In this section we return to the spirit of the last chapter. Two new concepts will be introduced in abstract projective space. They will then be connected with our coordinatization scheme. Both concepts derive from the following definition.

DEFINITION 1 A *complete quadrangle* in a projective space is a set of four coplanar points, no three collinear, together with the six lines joining them (Figure IV.3.1).

---

* Exercises on frames of reference will appear mainly as applications in Chapter V.

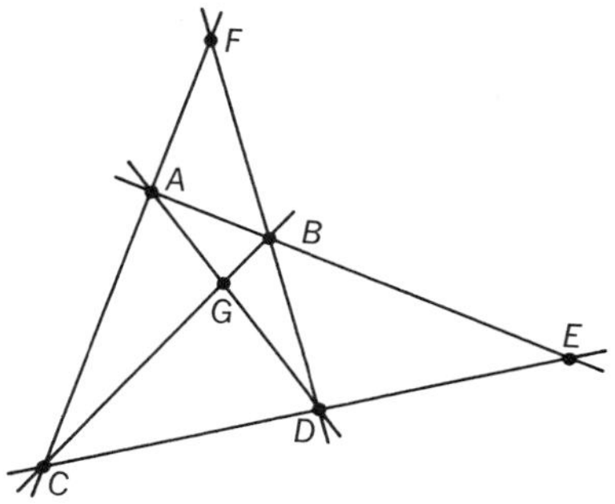

Fig. IV.3.1

The four points A, B, C, D of a complete quadrangle are called the *vertices;* the joining lines are the *edges*. The points $E = AB \wedge CD$, $F = AC \wedge BD$, $G = AD \wedge BC$ are the *diagonal points* and the lines EF, EG, FG the *diagonals* of the quadrangle. It is easy to see that the diagonal points of any complete quadrangle are distinct. This is not necessarily true of the diagonals. In fact, the Fano plane (Figure II.3.1) is (in several different ways) a complete quadrilateral with only one diagonal. This suggests the next definition.

DEFINITION 2 A projective space is *Fano* if the diagonal points of every complete quadrilateral in the space are collinear.*

The analytic interpretation of this is as follows:

THEOREM 1 If one plane in a projective space is Fano, then every plane in the space is Fano. A projective plane is Fano if and only if in every planar ternary ring of coordinates $1 + 1 = 0$.

PROOF. The first statement is a consequence of the isomorphism between any two planes in a projective space.

For the second statement notice that the vertices of an arbitrary complete quadrangle in a projective plane (Figure IV.3.1) can be given the coordinates $A⋮(1, 0)$, $B⋮(0, 1)$, $C⋮(0)$, $D⋮(\infty)$. Then the diagonal points would have coordinates $E⋮(1)$, $F⋮(0, 1)$, $G⋮(1, 1)$. We are to show that these points are collinear if and only if $1 + 1 = 0$. But this is immediate from the definition of addition (III.1.2). ∎

For the planes of this chapter we immediately have a corollary.

COROLLARY The coordinatizing field of a Pappian projective plane (or of a projective space in which one plane is Pappian) is of characteristic two if and only if the plane (or space) is Fano.

---

* The term should really be *anti-Fano*, since Fano's axiom excluded such quadrilaterals. But the above term is prevalent.

For more on Fano planes, see Section VI.5. We turn now to the second concept.

DEFINITION 3 An ordered quadruple of points (A, B, C, D) in a projective space forms a *harmonic tetrad* if there exists a complete quadrangle such that:

(1) A and B are vertices of the quadrangle;
(2) C is the (necessarily unique) diagonal point of the quadrangle that is on AB;
(3) D is the intersection of AB with the diagonal of the quadrangle that contains the other (than C) diagonal points (Figure IV.3.2).

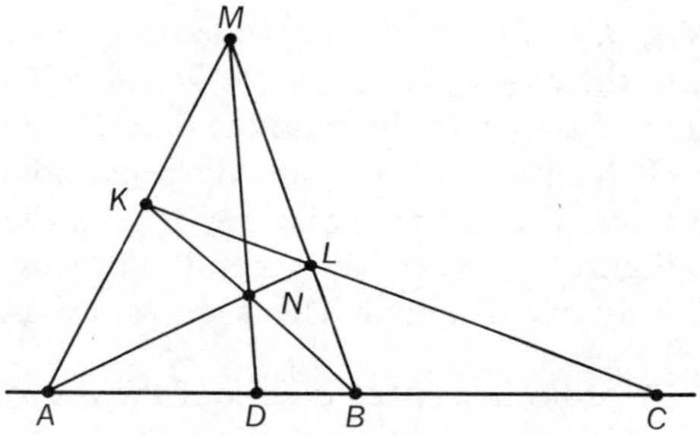

Fig. IV.3.2

In the figure the complete quadrangle has vertices A, B, K, L. If (A, B, C, D) form a harmonic tetrad we write $H(AB, CD)$ and sometimes say that A and B *separate harmonically* C and D. (More on this is given in Section 6.) Obviously, $H(AB, CD)$ implies: A, B, C, D collinear; A, B, C distinct; A, B, D distinct. In general, $C = D$ if and only if the plane of Figure IV.3.2 is Fano. Incidentally, if Figure IV.3.2 can be found at all, it can clearly be found in every plane containing AB. Thus, we shall study this concept only in the plane.

Referring again to Figure IV.3.2, quadrangle A, B, K, L also shows that $H(BA, CD)$. Quadrangle A, B, M, N shows that $H(AB, DC)$. Thus, at most six permutations of (A, B, C, D) can make any difference in this concept. This suggests (via the corollary to Theorem 2.4) a connection with cross ratio, which we now establish.

THEOREM 2 In a Pappian projective plane, $H(AB, CD)$ if and only if in the associated homogeneous coordinate system

$$(ABCD) = -1.$$

PROOF. In Figure IV.3.2 we may assume $AB{:}[0, 1, 0]$, $A{:}(0, 0, 1)$, $B{:}(1, 0, 1)$, $C{:}(1, 0, 2)$, $AK{:}[1, 0, 0]$, $K{:}(0, 1, 1)$, $M{:}(0, 1, 0)$. Then, by easy calculations, KC:$[2, 1, -1]$, MB:$[1, 0, -1]$, L:$(1, -1, 1)$, AL:$[1, 1, 0]$, BK:$[1, 1, -1]$, N:$(1, -1, 0)$, MN:$[0, 0, 1]$, and $AB \wedge MN$:$(1, 0, 0)$.

Now, if $H(AB, CD)$, then D:$(1, 0, 0) = (-1)(0, 0, 1) + (1, 0, 1)$, so $(ABCD) = -1$. Conversely, if $(ABCD) = -1$, then D:$(-1)(0, 0, 1) + (1, 0, 1) = (1, 0, 0)$, and so $D = AB \wedge MN$, whence $H(AB, CD)$. ❙

Theorem 2.3 immediately yields a corollary.

COROLLARY If A, B, C are distinct collinear points in a Pappian projective plane, there is one and only one point D such that $H(AB, CD)$.

Exercises

*1. Prove the corollary to Theorem 2 without any use of coordinates. (Hint: Use Exercise III.2.7.)

2. By Theorem 2 and the corollary to Theorem 2.4, $H(AB, CD)$ in a Pappian projective plane implies $H(CD, AB)$. Prove this in as general a projective plane as possible without any use of coordinates.

*3. Find a sufficient condition for the coordinatizing field of a Pappian plane to be of characteristic three.

4. Dualize the definitions and results of this section.

*5. (H. S. M. Coxeter.) Prove, in a non-Fano Desarguesian projective plane, that the six edges of a quadrangle meet the three sides of its diagonal triangle in the six vertices of a quadrilateral that has the same diagonal triangle.

4. THE FUNDAMENTAL THEOREM

The principal tools of this section are the cross ratio of four points, the dual concept of the cross ratio of four lines, and a theorem connecting these two concepts.

Let $\mathbf{P}^1$, $\mathbf{P}^2$, $\mathbf{P}^3$, $\mathbf{P}^4$ be four distinct hyperplanes of a one-dimensional bundle in some subspace of $\mathbf{P}_nF$ (see Section 1 if necessary). Choose hyperplane coordinate vectors $\mathbf{P}^i$:$\mathbf{u}_i$, $1 \leq i \leq 4$, such that $\mathbf{u}_3 = \mathbf{u}_1 + \mathbf{u}_2$. Then $\mathbf{u}_4 = y_1\mathbf{u}_1 + y_2\mathbf{u}_2$ for some nonzero y_1 and y_2 in F. By the dual (of the one-dimensional case) of Theorem 2.1, the number $y_1y_2^{-1}$ is uniquely determined by the ordered quadruple $(\mathbf{P}^1, \mathbf{P}^2, \mathbf{P}^3, \mathbf{P}^4)$. Define the *cross ratio*

$$(\mathbf{P}^1\mathbf{P}^2\mathbf{P}^3\mathbf{P}^4) = y_1y_2^{-1},$$

and for $\mathbf{P}^1 = \mathbf{P}^4$ $(y_2 = 0)$ define it as ∞.

By duality all previous properties of cross ratio hold. The theorem connecting cross ratio of points and of hyperplanes is as follows.

THEOREM 1 Let $\mathbf{P}^1$, $\mathbf{P}^2$, $\mathbf{P}^3$, $\mathbf{P}^4$ be hyperplanes in a one-dimensional bundle in a subspace $\mathbf{L}$ of $\mathbf{P}_nF$. (The hyperplanes are assumed distinct except possibly $\mathbf{P}^1 = \mathbf{P}^4$.) Let l be a line in $\mathbf{L}$ not through the center of the bundle. Let $P_1 = \mathbf{P}^1 \wedge l$, $P_2 = \mathbf{P}^2 \wedge l$, $P_3 = \mathbf{P}^3 \wedge l$. Then a point P_4 is $\mathbf{P}^4 \wedge l$ if and only if*

$$(P_1P_2P_3P_4) = (\mathbf{P}^1\mathbf{P}^2\mathbf{P}^3\mathbf{P}^4). \tag{1}$$

PROOF. $P_4 = P_1$ iff $\mathbf{P}^4 = \mathbf{P}^1$ iff (1) holds and equals ∞. Now assume $P_4 \neq P_1$, $\mathbf{P}^4 \neq \mathbf{P}^1$. Then choose coordinate vectors $P_i \colon \mathbf{p}_i$ and $\mathbf{P}^i \colon \mathbf{u}_i$ such that $\mathbf{p}_3 = \mathbf{p}_1 + \mathbf{p}_2$, $\mathbf{p}_4 = x\mathbf{p}_1 + \mathbf{p}_2$, $\mathbf{u}_3 = \mathbf{u}_1 + \mathbf{u}_2$, and $\mathbf{u}_4 = y\mathbf{u}_1 + \mathbf{u}_2$. By hypothesis $\mathbf{u}_i \cdot \mathbf{p}_i = 0$, $i = 1, 2, 3$. Then

$$0 = \mathbf{u}_3 \cdot \mathbf{p}_3 = (\mathbf{u}_1 + \mathbf{u}_2)\cdot(\mathbf{p}_1 + \mathbf{p}_2) = \mathbf{u}_1 \cdot \mathbf{p}_2 + \mathbf{u}_2 \cdot \mathbf{p}_1. \tag{*}$$

ASSERTION. $\mathbf{u}_1 \cdot \mathbf{p}_2 \neq 0$. For if $\mathbf{u}_1 \cdot \mathbf{p}_2 = 0$ then P_1 and P_2 would lie on $\mathbf{P}^1$, so l would lie on $\mathbf{P}^1$, a contradiction (why?).

Now P_4 is $l \wedge \mathbf{P}^4$ iff $\mathbf{u}_4 \cdot \mathbf{p}_4 = 0$, iff $0 = (y\mathbf{u}_1 + \mathbf{u}_2)\cdot(x\mathbf{p}_1 + \mathbf{p}_2) = y\mathbf{u}_1 \cdot \mathbf{p}_2 + x\mathbf{u}_2 \cdot \mathbf{p}_1 \overset{(*)}{=} (y - x)\mathbf{u}_1 \cdot \mathbf{p}_2$, iff (by assertion) $y = x$, that is, iff (1). ∎

We can now show how cross ratio provides a connection between isomorphisms in $\mathbf{P}_nF$ and the algebraic structure of F.

THEOREM 2 Let μ be an isomorphism from one r-dimensional subspace $\mathbf{P}_r$ of $\mathbf{P}_nF$ to another, where $r > 1$. Then there is an automorphism α_μ of F such that

$$(P_1\mu P_2\mu P_3\mu P_4\mu) = (P_1P_2P_3P_4)\alpha_\mu, \tag{2}$$

whenever P_1, P_2, P_3, P_4 are distinct collinear points in P_r.

PROOF. Pick a line l in P_r. Let $(Q_0Q_1|U)$ be a frame on l. Say P on l, $P \neq Q_0$, has affine coordinate x. Then $P\mu$ is on $l\mu$, $P\mu \neq Q_0\mu$, so in the frame $(Q_0\mu Q_1\mu|U\mu)$ $P\mu$ has an affine coordinate, say, x'. Define $x\alpha_\mu = x'$. α_μ is one–one and onto because μ, restricted to $\tilde{l} - \{Q_0\}$, is. Since $r > 1$ there is a plane $\mathbf{P}_2$ in $\mathbf{P}_r$ containing l. The configurations connecting points on l with affine coordinates a, b, $a + b$, and ab (Figure III.5.5) are preserved by μ. Hence, α_μ preserves addition and multiplication. Therefore, α_μ is an automorphism of F.

Let P_1, P_2, P_3, P_4 be distinct points on l, with $Q_0 \neq P_i$, $1 \leq i \leq 4$. Say P_i has affine coordinate x_i. Then $P_i\mu$ has affine coordinate $x_i\alpha_\mu$ and, by Theorem 2.4, $(P_1\mu P_2\mu P_3\mu P_4\mu) = (x_1\alpha_\mu - x_3\alpha_\mu)(x_2\alpha_\mu - x_4\alpha_\mu)/(x_1\alpha_\mu - x_4\alpha_\mu)(x_2\alpha_\mu - x_3\alpha_\mu) = [(x_1 - x_3)(x_2 - x_4)/(x_1 - x_4)(x_2 - x_3)]\alpha_\mu$ (since α_μ is an automorphism) $= (P_1P_2P_3P_4)\alpha_\mu$. To show that (2) holds generally we must show: (a) the definition of α_μ is independent of choice of Q_0, Q_1, U on l; (b) the definition of α_μ is independent of choice of l in $\mathbf{P}_r$.

---

* The reader should convince himself that these four points exist and are unique.

(a) Given x in F find $P_1, P_2, P_3, P_4 \neq Q_0$ on l such that $(P_1P_2P_3P_4) = x$. (This can clearly be done for all but at most one x.) Then $x\alpha_\mu = (P_1\mu P_2\mu P_3\mu P_4\mu)$, so α_μ can be defined (for all but at most one x, hence for all x) without reference to Q_0, Q_1, U.

(b) Let m be a different line in $\mathbf{P}_r$. Suppose α'_μ is obtained, as above, from m. Let $x \in F$. We must show $x\alpha_\mu = x\alpha'_\mu$.

CASE 1. l intersects m. Let $\mathbf{P}_2$ be the plane on l and m. Let V be a point of P_2 not on l or m. Choose B_i on m such that $(B_1B_2B_3B_4) = x$. Let $A_i = l \wedge VB_i$, $1 \leq i \leq 4$. By Theorem 1, $x = (B_1B_2B_3B_4) = (VB_1VB_2VB_3VB_4) = (A_1A_2A_3A_4) = x$. Now μ preserves incidence, so in the image space we have, again by Theorem 1, $x\alpha'_\mu = (B_1\mu B_2\mu B_3\mu B_4\mu) = (A_1\mu A_2\mu A_3\mu A_4\mu) = x\alpha_\mu$.

CASE 2. l does not intersect m. Pick any points A on l and B on m. Let $n = AB$. Let α''_μ be the automorphism obtained from n. Then applying Case 1 twice $x\alpha'_\mu = x\alpha''_\mu = x\alpha_\mu$. ▮

COROLLARY Under the hypotheses of the theorem,

$$(\mathbf{P}^1\mu\mathbf{P}^2\mu\mathbf{P}^3\mu\mathbf{P}^4\mu) = (\mathbf{P}^1\mathbf{P}^2\mathbf{P}^3\mathbf{P}^4)\alpha_\mu, \tag{3}$$

whenever $\mathbf{P}^1$, $\mathbf{P}^2$, $\mathbf{P}^3$, $\mathbf{P}^4$ are hyperplanes in a one-dimensional bundle in $\mathbf{P}_r$.

PROOF. The proof is clear from (2) and Theorem 1. ▮

The case $r = 1$, excluded above, requires a different proof. First, notice that Figure IV.3.2 is preserved by isomorphisms. Thus, every isomorphism on a space of dimension greater than one preserves harmonic tetrads. This provides motivation for an addition to the definition of isomorphism.

ADDITION TO DEFINITION II.5.1 An isomorphism on a one-dimensional projective space is assumed to preserve harmonic tetrads.

With this we can extend Theorem 2.

THEOREM 2′ The previous theorem remains valid for $r = 1$, provided F is not of characteristic two.

PROOF. Here we are dealing with an isomorphism μ from a line l to a line $l' = l\mu$ in $\mathbf{P}_nF$, $n \geq 1$. Let $(Q_0Q_1|U)$ be a frame on l. Choose coordinate vectors satisfying $\mathbf{q}_0 + \mathbf{q}_1 = \mathbf{u}$. Denote by P_x the point on l with coordinate vector $\mathbf{p}_x = x\mathbf{q}_0 + \mathbf{q}_1$. Then $Q_1 = P_0$, $U = P_1$. Let $P_\infty = Q_0$. Now, decorating images of μ with primes, do the same thing on l' with the frame $(P'_\infty P'_0|P'_1)$. Define α_μ (just as before) by

$$P_x\mu = P'_{x\alpha_\mu}.$$

Then, by definition, $0\alpha_\mu = 0$ and $1\alpha_\mu = 1$. α_μ is clearly one–one and onto. It remains only to show that α_μ preserves addition and multiplication.

Let $a, b \in F$, $a \neq b$. Then $\mathbf{p}_a + \mathbf{p}_b = (a + b)\mathbf{p}_\infty + 2\mathbf{p}_0$, which (since F is not of characteristic two!) is a coordinate vector for $P_{a+b/2}$, and $(-1)\mathbf{p}_a + \mathbf{p}_b = (b - a)\mathbf{p}_\infty$, which is a coordinate vector for P_∞. Hence, $(P_a P_b P_{a+b/2} P_\infty) = -1$, so, by Theorem 3.2,

$$H(P_a P_b,\ P_{a+b/2} P_\infty).$$

Then, by definition,

$$H(P'_{a\alpha_\mu} P'_{b\alpha_\mu},\ P'_{(a+b/2)\alpha_\mu} P'_\infty),$$

so, by Theorem 3.2,

$$(P'_{a\alpha_\mu} P'_{b\alpha_\mu} P'_{(a+b/2)\alpha_\mu} P'_\infty) = -1. \tag{i}$$

But

$$\mathbf{p}'_{a\alpha_\mu} + \mathbf{p}'_{b\alpha_\mu} = (a\alpha_\mu + b\alpha_\mu)\mathbf{p}'_\infty + 2\mathbf{p}'_0$$

and

$$(-1)\mathbf{p}'_{a\alpha_\mu} + \mathbf{p}'_{b\alpha_\mu} = (b\alpha_\mu - a\alpha_\mu)\mathbf{p}'_\infty.$$

So, as before,

$$(P'_{a\alpha_\mu} P'_{b\alpha_\mu} P'_{a\alpha_\mu + b\alpha_\mu/2} P'_\infty) = -1. \tag{ii}$$

Comparing (i) and (ii), we have, by Theorem 2.3, the corollary to Theorem 2.4, and the uniqueness of affine coordinates

$$\left(\frac{a+b}{2}\right)\alpha_\mu = \frac{a\alpha_\mu + b\alpha_\mu}{2}. \tag{iii)*}$$

Note that (iii) holds for all a, b, including $a = b$. Setting $b = 0$ in (iii):

$$\left(\frac{a}{2}\right)\alpha_\mu = \frac{a\alpha_\mu}{2}. \tag{iv}$$

Then

$$(a+b)\alpha_\mu \overset{\text{(iv)}}{=} 2\left(\frac{a+b}{2}\right)\alpha_\mu \overset{\text{(iii)}}{=} a\alpha_\mu + b\alpha_\mu,$$

that is, α_μ preserves addition (hence, also subtraction).

Now let $a, b \in F$, $a \neq b$, $a \neq -b$. Then

$$b\mathbf{p}_a + a\mathbf{p}_b = 2ab\mathbf{p}_\infty + (a+b)\mathbf{p}_0,$$

which is a coordinate vector for $P_{2ab/(a+b)}$, and

$$(-1)(b\mathbf{p}_a) + a\mathbf{p}_b = (a - b)\mathbf{p}_\infty,$$

which is a coordinate vector for P_∞. Thus,

$$(P_a P_b P_{(2ab/a+b)} P_\infty) = -1.$$

As before, we also have

$$(P'_{a\alpha_\mu} P'_{b\alpha_\mu} P_{(2ab/a+b)\alpha_\mu} P'_\infty) = (P'_{a\alpha_\mu} P'_{b\alpha_\mu} P'_{2a\alpha_\mu b\alpha_\mu/a\alpha_\mu + b\alpha_\mu} P'_\infty) = -1,$$

* That is, α_μ preserves arithmetic means.

and, therefore,

$$\left(\frac{2ab}{a+b}\right)\alpha_\mu = \frac{2a\alpha_\mu b\alpha_\mu}{a\alpha_\mu + b\alpha_\mu}.$$

Then, by (iii),

$$\left(\frac{ab}{a+b}\right)\alpha_\mu = \frac{a\alpha_\mu b\alpha_\mu}{a\alpha_\mu + b\alpha_\mu}. \qquad \text{(v)*}$$

Note that (v) also holds for $a = b$. Let c be in F. Let $a = 1 + c$, $b = 1 - c$. Then by (v) ($a \neq -b$ since F is not of characteristic two!) and (iv)

$$\left(\frac{ab}{a+b}\right)\alpha_\mu = \left(\frac{1-c^2}{2}\right)\alpha_\mu = \frac{1}{2}(1 - (c^2)\alpha_\mu) \overset{(v)}{=} \frac{a\alpha_\mu b\alpha_\mu}{a\alpha_\mu + b\alpha_\mu}$$

$$= \frac{(1 + c\alpha_\mu)(1 - c\alpha_\mu)}{2},$$

so that

$$1 - (c^2)\alpha_\mu = (1 + c\alpha_\mu)(1 - c\alpha_\mu) = 1 - (c\alpha_\mu)^2,$$

and

$$(c^2)\alpha_\mu = (c\alpha_\mu)^2. \qquad \text{(vi)}$$

Finally, for any $a, b \in F$,

$$(ab)\alpha_\mu = [\tfrac{1}{2}((a+b)^2 - a^2 - b^2)]\alpha_\mu \underset{(iii)}{\overset{(iv)}{=}} \tfrac{1}{2}[((a+b)^2)\alpha_\mu - (a^2)\alpha_\mu - (b^2)\alpha_\mu]$$

$$\overset{(vi)}{=} \tfrac{1}{2}[(a\alpha_\mu + b\alpha_\mu)^2 - (a\alpha_\mu)^2 - (b\alpha_\mu)^2] = a\alpha_\mu b\alpha_\mu. \quad \blacksquare$$

We come now to the main result of this chapter.

THEOREM 3 (The "fundamental theorem of projective geometry.") Let $(Q_0Q_1 \cdots Q_r \mid U)$, $(Q_0'Q_1' \cdots Q_r' \mid U')$ be frames of reference for r-dimensional subspaces $\mathbf{P}_r$, $\mathbf{P}_r'$ respectively of $\mathbf{P}_nF$. Let α be an automorphism of F. Then there exists a unique isomorphism μ from $\mathbf{P}_r$ onto $\mathbf{P}_r'$ such that $Q_i\mu = Q_i'$, $0 \leq i \leq r$, $U\mu = U'$, and $\alpha_\mu = \alpha$.

PROOF

Existence: If P in $\mathbf{P}_r$ has coordinates $(x_0, x_1, \cdots, x_r)$ in the frame $(Q_0Q_1 \cdots Q_r|U)$, let $P\mu$ be the point in $\mathbf{P}_r'$ with coordinates $(x_0\alpha, x_1\alpha, \cdots, x_r\alpha)$ in the frame $(Q_0'Q_1' \cdots Q_r' \mid U')$. If $t \neq 0$ then $t\alpha \neq 0$ and $((tx_0)\alpha, (tx_1)\alpha, \cdots, (tx_r)\alpha) = t\alpha(x_0\alpha, x_1\alpha, \cdots, x_r\alpha)$. Thus, μ is defined independently of the representative coordinate vectors. Since α is one–one and onto, so is μ. Since α is an automorphism, μ preserves linear combinations. Thus, μ is an isomorphism. Since $1\alpha = 1$ for any automorphism, we have $Q_i\mu = Q_i'$, $0 \leq i \leq r$, and $U\mu = U'$. Now let $x \in F$. Find points P_1, P_2, P_3, P_4 in $\mathbf{P}_r$ such

---

* That is, α_μ preserves harmonic means.

that $(P_1P_2P_3P_4) = x$. Then there are coordinate vectors satisfying $\mathbf{p}_1 + \mathbf{p}_2 = \mathbf{p}_3$, $\mathbf{p}_4 = x\mathbf{p}_1 + \mathbf{p}_2$. Applying* α: $\mathbf{p}_1\alpha + \mathbf{p}_2\alpha = \mathbf{p}_3\alpha$, $\mathbf{p}_4\alpha = (x\alpha)\mathbf{p}_1\alpha + \mathbf{p}_2\alpha$, so, by definition of μ,

$$(P_1\mu P_2\mu P_3\mu P_4\mu) = x\alpha.$$

Then, by Eq. (2), $x\alpha = x\alpha_\mu$ for all x, that is, $\alpha = \alpha_\mu$.

Uniqueness:† Let μ^* be another isomorphism enjoying the same properties as μ. We wish to show that $\nu = \mu\mu^{*-1}$ is the identity function. By Exercise 2 ν is a collineation of $\mathbf{P}_r$ and $\alpha_\nu = \alpha_\mu\alpha_{\mu^*}^{-1} = 1$, the identity automorphism, since we are assuming $\alpha_\mu = \alpha_{\mu}^*$.

ASSERTION 1. If ν fixes three distinct points on a line, it fixes every point on the line. Let P_1, P_2, P_3 be distinct points on the line l such that $P_i\nu = P_i$, $i = 1, 2, 3$. Let P be any other point on l. Then, by Eq. (2), since $\alpha_\nu = 1$, $(P_1P_2P_3P\nu) = (P_1P_2P_3P)$, so, by Theorem 2.3, $P\nu = P$.

ASSERTION 2. The subspace spanned by $Q_0, Q_1, \cdots, Q_s$, is fixed pointwise by ν for each $s \leq r$. We prove this by induction on s. The assertion is trivially true for $s = 0$. Suppose it true for $s - 1$, where $s \geq 1$. Each point in the sets $\{Q_0, Q_1, \cdots, Q_s\}$ and $\{Q_{s+1}, \cdots, Q_r, U\}$ is fixed by ν. Therefore, it fixes *setwise* the subspaces $[Q_0, Q_1, \cdots, Q_s]$ and $[Q_{s+1}, \cdots, Q_r, U]$. Hence, it fixes setwise the space $\mathbf{V} = [Q_0, Q_1, \cdots, Q_s] \wedge [Q_{s+1}, \cdots, Q_r, U]$. But $\mathbf{V}$ is of dimension $r - s - (r - s) = 0$, that is, $\mathbf{V}$ consists of a single point, say, V_s, which is fixed by ν. The point $P_s = Q_sV_s \wedge [Q_0, Q_1, \cdots, Q_{s-1}]$ exists $(s - 1 - (s - 1) = 0)$, and, since it lies on $[Q_0, Q_1, \cdots, Q_{s-1}]$, it is fixed by ν. Thus ν fixes P_s, V_s, Q_s and so, by Assertion 1, ν fixes every point on V_sQ_s. Now let P be any point in $[Q_0, Q_1, \cdots, Q_s]$ not on $[Q_0, Q_1, \cdots, Q_{s-1}]$ or Q_sV_s. The plane $[P, Q_s, V_s]$ meets $[Q_0, Q_1, \cdots, Q_{s-1}]$ in a line l $(s - 1 + 2 - s = 1)$. Let m_1 and m_2 be distinct lines in $[P, Q_s, V_s]$ through P. $m_i \wedge V_sQ_s$ and $m_i \wedge l$ are fixed by ν, hence, $m_i\nu = m_i$, $i = 1, 2$, and so $P\nu = P$. Thus, ν fixes $[Q_0, Q_1, \cdots, Q_s]$ pointwise, and the induction is complete.

The case $s = r$ of Assertion 2 is the desired result. ▎

We close this section by determining all collineations of an analytic projective space.

THEOREM 4 Let μ be a collineation of $\mathbf{P}_nF$. Then there is a unique automorphism α_μ of F and an $(n + 1) \times (n + 1)$ nonsingular matrix A_μ over F, unique up to a nonzero scalar multiple, such that if a point P:$[\mathbf{x}]$ then

$$P\mu\colon[A_\mu\mathbf{x}\alpha_\mu]. \tag{4}$$ ‡

---

* If $\mathbf{x} = (x_0, x_1, \cdots, x_r) \in F^{r+1}$ and α is a function on F, we denote $(x_0\alpha, x_1\alpha, \cdots, x_r\alpha)$ by $\mathbf{x}\alpha$.

† Here if $r = 1$ we assume char $F \neq 2$.

‡ The notation $A\mathbf{x}\alpha$ is ambiguous. We shall *always* interpret it as $A(\mathbf{x}\alpha)$.

Conversely, every such automorphism and matrix determines, by (4), a collineation of $\mathbf{P}_nF$.

PROOF. Let μ be a collineation of $\mathbf{P}_nF$. Let α_μ be the α_μ of Theorem 2 (or 2′). If a point $P{:}\mathbf{x}$ then $P{:}\mathbf{x}$ in the "natural frame" $(Q_0Q_1 \cdots Q_n|U)$, where $Q_0 = [(1, 0, \cdots, 0)]$, $Q_1 = [(0, 1, 0, \cdots, 0)]$, $\cdots$, $Q_n = [(0, \cdots, 0, 1)]$, $U = [(1, 1, \cdots, 1)]$. Then $P\mu{:}\mathbf{x}\alpha_\mu$ in the image frame $(Q_0\mu Q_1\mu \cdots Q_n\mu|U\mu)$. Say $P\mu{:}\mathbf{x}'$ in $\mathbf{P}_nF$. Then, regarding $(Q_0\mu Q_1\mu \cdots Q_n\mu|U_\mu)$ as the "old" frame and the natural frame as the "new" frame, the matrix A_μ of (4) is the matrix A of Theorem 2.2. The converse is trivial. ∎

In the general case, where F may be only a division ring, the statement of the uniqueness of the automorphism α_μ must be replaced. Let a be a nonzero element of a division ring D. The function α_a: $D \to D$ defined by $x\alpha_a = axa^{-1}$ is an automorphism of D, called an *inner automorphism*. Clearly, D is a field iff the only inner automorphism of D is the identity automorphism. In general, the automorphism of Theorem 4 is only unique up to an inner automorphism. More precisely: Two automorphisms α and β of a division ring D belong to the same collineation of $\mathbf{P}_nD$ if and only if $\alpha\beta^{-1}$ is an inner automorphism of D. For a proof of this, and for much more information on the collineations of projective spaces, see E. Artin's *Geometric Algebra*.

In a vector space V over a field F a transformation $\beta : \mathbf{x} \to A\mathbf{x}\alpha$, where A is a linear transformation and α is an automorphism, has the property

$$(a\mathbf{x} + b\mathbf{y})\beta = (a\alpha)\mathbf{x}\beta + (b\alpha)\mathbf{y}\beta \tag{5}$$

for all $\mathbf{x}, \mathbf{y} \in V$, $a, b \in F$. Such transformations are called *semilinear*. Thus, collineations of $\mathbf{P}_nF$ correspond to classes of proportional nonsingular semilinear transformations of F^{n+1}. To single out the more special case of *linear* transformations we have a definition and a corollary.

DEFINITION 1 An isomorphism defined on a subspace of $\mathbf{P}_nF$ is *projective* if its associated automorphism (Theorems 2 and 2′) is the identity function.

COROLLARY Every collineation of $\mathbf{P}_nF$, where F is a field having no nonidentity automorphisms (such as the real field), is projective.

PROOF. Proof is immediate from Theorem 4. ∎

Projective collineations are often called *projectivities*.

Exercises

1. If μ is an isomorphism from $\mathbf{P}_r$ to $\mathbf{P}'_r$, show that μ^{-1} is an isomorphism from $\mathbf{P}'_r$ to $\mathbf{P}_r$ and that $\alpha_{\mu^{-1}} = \alpha_\mu^{-1}$.

2. If μ_1 is an isomorphism from $\mathbf{P}_r^1$ to $\mathbf{P}_r^2$ and μ_2 is an isomorphism from $\mathbf{P}_r^2$ to $\mathbf{P}_r^3$, show that $\mu_1\mu_2$ is an isomorphism from $\mathbf{P}_r^1$ to $\mathbf{P}_r^3$ and that $\alpha_{\mu_1\mu_2} = \alpha_{\mu_1}\alpha_{\mu_2}$. State the results of this and the previous exercise in a nice algebraic way in the case that $\mathbf{P}_r$, $\mathbf{P}_r'$, $\mathbf{P}_r^i$ are the same space.
3. Suppose a collineation μ of $\mathbf{P}_nF$ has the representation (4). Suppose that in a new frame of reference a point P:$\mathbf{x}$ has coordinates $\mathbf{y}$. Prove that there exists a nonsingular matrix B, independent of P, such that the representation of μ in the new frame is
$$[\mathbf{y}'] = [BA(B\alpha_\mu)^{-1}\mathbf{y}\alpha_\mu].$$
It will be necessary first to prove $B\alpha_\mu$ nonsingular.* (Collineations that differ only by a change of frame are called *equivalent*. The classification of projective collineations under equivalence is then just the classification of nonsingular matrices under similarity.)
4. By proper selection of the frame of reference, show that every collineation of $\mathbf{P}_3F$ has a matrix with more than one half the entries zero.
5. Show that every projectivity from a line l to a line l' in $\mathbf{P}_2F$ is the product of at most three projections (see Exercise II.1.3). Generalize to higher dimensions.

5. CORRELATIONS

There is an obvious dual theory to the theory of collineations. One begins with a one–one map of the set of all hyperplanes in one space to the set of all hyperplanes in another space and ends with Theorem 4.4 expressed in hyperplane coordinates. In this section we shall combine the theory of collineations with the dual theory in such a way as to obtain a new type of mapping of projective spaces, which is sometimes useful.

* * *

In order to get more quickly to the main results, we shall assume more than necessary in the basic definition. (See Exercise 1.)

DEFINITION 1 A *correlation* from a projective space $\mathbf{S}$ to a projective space $\mathbf{S}'$ is a one–one function from the set of all points in $\mathbf{S}$ onto the set of all hyperplanes in $\mathbf{S}'$ that sends dependent (independent) points in $\mathbf{S}$ to dependent (respectively, independent) hyperplanes in $\mathbf{S}'$. (If $\dim \mathbf{S} = 1$, it is further assumed that harmonic tetrads are preserved.)

Thus, a correlation sends an r-dimensional subspace of $\mathbf{S}$ onto an r-dimensional bundle in $\mathbf{S}'$. The simplest, as well as the most important, example of a correlation is the *duality correlation* δ of $\mathbf{P}_nF$:

---

* By $B\alpha$ we mean $(b_{ij}\alpha)$, where $B = (b_{ij})$.

$$[(x_0, x_1, \cdots, x_n)]\delta = [[x_0, x_1, \cdots, x_n]]. \tag{1}$$

We saw in Section 1 that this was a correlation. This allows us to extend previous results on collineations to results on correlations without difficulty.

THEOREM 1 Let κ be a correlation from an r-dimensional subspace $\mathbf{P}_r$ of $\mathbf{P}_nF$ onto another one $\mathbf{P}'_r$. Then there is an automorphism α_κ of F such that, whenever P_1, P_2, P_3, P_4 are distinct collinear points of $\mathbf{P}_r$,

$$(P_1\kappa P_2\kappa P_3\kappa P_4\kappa) = (P_1P_2P_3P_4)\alpha_\kappa. \tag{2}$$

Conversely, if α is an automorphism of F, $P_0, P_1, \cdots, P_{r+1}$ are points of $\mathbf{P}_r$, no $r + 1$ dependent, and $\mathbf{P}^0, \mathbf{P}^1, \cdots, \mathbf{P}^{r+1}$ are hyperplanes of $\mathbf{P}'_r$, no $r + 1$ dependent, then there exists a unique correlation κ from $\mathbf{P}_r$ onto $\mathbf{P}'_r$ such that $P_i\kappa = \mathbf{P}^i$, $0 \leq i \leq r + 1$, and $\alpha_\kappa = \alpha$.

PROOF. Establish a coordinate system in $\mathbf{P}'_r$ and let δ' be the corresponding duality correlation. Obviously, δ' preserves cross ratio. $\mu = \kappa\delta'^{-1}$ is a collineation from $\mathbf{P}_r$ to $\mathbf{P}'_r$. Let α_κ be the α_μ of Theorems 4.2 and 4.2′.

Now, let μ be the collineation sending P_i to $P'_i = \mathbf{P}^i\delta'^{-1}$, $0 \leq i \leq r + 1$, such that $\alpha_\mu = \alpha$ (Theorem 4.3). Then $\kappa = \mu\delta'$ fulfills the existence part of the converse.

Finally, suppose $\bar{\kappa}$ also fulfills the converse. Then the collineations $\mu = \kappa\delta'^{-1}$ and $\bar{\mu} = \bar{\kappa}\delta'^{-1}$ agree on the P_i and both induce α. Hence, by Theorem 4.3, $\mu = \bar{\mu}$ and so $\kappa = \bar{\kappa}$. ▮

With equal ease we may extend Theorem 4.

THEOREM 2 Let κ be a correlation from $\mathbf{P}_nF$ onto itself. Then there is a unique automorphism α_κ of F and a nonsingular $(n + 1) \times (n + 1)$ matrix A_κ over F, unique up to a nonzero scalar multiple, such that if a point $P\colon[\mathbf{x}]$ then

$$P\kappa\colon[A_\kappa\mathbf{x}\alpha_\kappa]. \tag{3}$$

PROOF. Exercise 2.

Recall that, by definition, a point $Q\colon\mathbf{y}$ is on a hyperplane $P\colon\mathbf{u}$ iff $\mathbf{y}\cdot\mathbf{u} = 0$. Thus, the correlation of (3) may be thought of as sending the point $P\colon\mathbf{x}$ to the hyperplane consisting of all points $Q\colon\mathbf{y}$ such that

$$\mathbf{y}\cdot A_\kappa\mathbf{x}\alpha_\kappa = 0. \tag{4}$$

This will often be the best way of interpreting the next definition.

DEFINITION 2 A correlation κ of $\mathbf{P}_nF$ with itself is *projective* if α_κ is the identity automorphism. A projective correlation is called a *polarity* if A_κ is symmetric, and a *nullity** if A_κ is skew-symmetric.

---

* Some authors say *null system*.

The geometrical significance of these special correlations is partly shown in the next result.

THEOREM 3 A correlation κ of $\mathbf{P}_nF$ with itself is a nullity if

$$P \in P\kappa \tag{5}$$

for all points P. The converse holds unless F is of characteristic two.

PROOF. Suppose κ is a correlation satisfying (5). Let $A = A_\kappa$, $\alpha = \alpha_\kappa$ (Theorem 2). Then $\mathbf{x}A\mathbf{x}\alpha = 0$ for all $\mathbf{x}$. Writing this out in terms of $A = (a_{ij})$, we have

$$\sum_{i,j} a_{ij}x_i(x_j\alpha) = 0, \tag{*}$$

for all x_i in F. For fixed i set $x_i = 1$, all other $x_j = 0$; then (*) becomes

$$a_{ii} = 0, \qquad 0 \leq i \leq n. \tag{**}$$

For fixed $i \neq k$ set $x_i = 1$, $x_k = x$, all other $x_j = 0$. Then from (*) and (**)

$$a_{ik}(x\alpha) + a_{ki}x = 0, \tag{†}$$

for all i, k and all x in F. In (†) set $x = 1$; then $a_{ik} = -a_{ki}$, so A is skew-symmetric. Now A is nonsingular, so there exists $i \neq k$ such that $a_{ik} = -a_{ki} \neq 0$. For this choice (†) becomes

$$x\alpha - x = 0$$

for all x. Hence, α is the identity, and κ is a nullity.

Conversely, if κ is a nullity let $A = A_\kappa$. Let $P\colon\mathbf{x}$ be any point. Then $P\kappa\colon A\mathbf{x}$ and $\mathbf{x}A\mathbf{x} = (\mathbf{x}A\mathbf{x})^T = \mathbf{x}A^T\mathbf{x} = -\mathbf{x}A\mathbf{x}$; so, if char $F \neq 2$, $\mathbf{x}A\mathbf{x} = 0$, that is, $P \in P\kappa$. ∎

The exception of characteristic two is no surprise, since in this case nullities and polarities are the same.

Now a correlation κ from $\mathbf{S}$ to itself induces a "dual correlation" κ^* from the hyperplanes to the points of $\mathbf{S}$ as follows: If $\mathbf{H}$ is a hyperplane, then $\{P\kappa \mid P \in \mathbf{H}\}$ is a bundle of hyperplanes the center of which is a single point H. Define $H = \mathbf{H}\kappa^*$. κ^* could be called the *involute* of κ, in view of the next definition.

DEFINITION 3 A correlation κ of a projective space with itself is an *involution* (and is called *involutory*) if $\kappa\kappa^*$ is the identity function.

THEOREM 4 (R. Brauer.) Let κ be an involutory correlation of $\mathbf{P}_nF$ with matrix A and automorphism α. Then either $A^T = -A$ or $A^T = A\alpha$ (for some choice of A) and α^2 is the identity.

PROOF. First we find a matrix and automorphism for κ^*. Let $\mathbf{H}\colon\mathbf{u}$ be a hyperplane. Let $\mathbf{y} = A^{T-1}\mathbf{u}\alpha$. Then, for every point $P\colon\mathbf{x}$ on $\mathbf{H}$, $\mathbf{y}A\mathbf{x}\alpha =$

$(A^{T-1}\mathbf{u}\alpha)^T A\mathbf{x}\alpha = \mathbf{u}\alpha A^{-1}A\mathbf{x}\alpha = \mathbf{u}\alpha\mathbf{x}\alpha = (\mathbf{x}\cdot\mathbf{u})\alpha = 0$. Hence, point $C\colon\mathbf{y}$ is the center of the bundle $\{P\kappa \mid P \in \mathbf{H}\}$, that is, $C = \mathbf{H}\kappa^*$. Thus, a matrix for κ^* is A^{T-1} and the automorphism for κ^* is α.

Now if a point $P\colon\mathbf{x}$, then $P\kappa\kappa^*\colon A^{T-1}(A\mathbf{x}\alpha)\alpha = (A^{T-1}(A\alpha))\mathbf{x}\alpha^2$. If κ is an involution, then $P\kappa\kappa^* = P$ and so

$$B(x_0\alpha^2, x_1\alpha^2, \cdots, x_n\alpha^2) = t(x_0, x_1, \cdots, x_n), \tag{*}$$

for some $t \neq 0$, where $B = A^{T-1}(A\alpha)$. For fixed j set $x_j = 1$ and all other $x_i = 0$ in (*) to obtain $B\mathbf{e}_j = t\mathbf{e}_j$, where $\mathbf{e}_j = (0, \cdots, 1, \cdots, 0)$, with the 1 in the $(j+1)$st place. This holds for $0 \leq j \leq n$; hence, $B = tI$, where I is the identity matrix. Thus, (*) becomes $t(x_i\alpha^2) = tx_i$, $0 \leq i \leq n$. In particular, $x\alpha^2 = x$ for all x in F, that is, α^2 is the identity.

Now $B = A^{T-1}(A\alpha) = tI$, so

$$A\alpha = tA^T. \tag{**}$$

Applying α to (**),

$$A = A\alpha^2 = t\alpha(A\alpha)^T = t\alpha(tA^T)^T = t(t\alpha)A.$$

Therefore,

$$t(t\alpha) = 1 \tag{†}$$

CASE 1. $t \neq -1$. Then $1 + t \neq 0$. Let $r = (1 + t)^{-1}$. Then

$$r\alpha = ((1 + t)\alpha)^{-1} = (1 + t\alpha)^{-1} \overset{(\dagger)}{=} t(1 + t)^{-1} = tr.$$

Now rA is also a matrix for κ. Replacing A in (**) by rA, we have $(rA)\alpha = t(rA)^T$, that is, $r\alpha A\alpha = trA^T$. But we have shown $r\alpha = tr$; hence,

$$A\alpha = A^T. \tag{††}$$

CASE 2. $t = -1$ and α is not the identity. Then $s\alpha \neq s$ for some s in F. Let $r = s - s\alpha$. Then $r \neq 0$ and $r\alpha = -r = rt$. Again replacing A in (**) by rA yields (††).

CASE 3. α is the identity. Then (†) becomes $t^2 = 1$, so (**) becomes either $A = -A^T$ or $A = A^T$ $(= A\alpha)$. ▎

COROLLARY A projective involutory correlation of $\mathbf{P}_nF$ is either a polarity or a nullity.

PROOF. Immediate from the theorem and Definition 2. ▎

Exercises

1. Assuming fewer properties for a correlation, derive Definition 1 as a theorem. (For motivation see definition and Theorem II.5.1.)
2. Prove Theorem 2.

3. Prove that if n is even then $\mathbf{P}_nF$ cannot have a nullity unless char $F = 2$.
4. Prove that a projective correlation of $\mathbf{P}_2F$ that maps each vertex of some triangle to the opposite side is a polarity.
5. Let P, A, B, C, A', B', C' be distinct points of $\mathbf{P}_2F$ with P, A, A' collinear; P, B, B' collinear; P, C, C' collinear; A, B, C not collinear. Find a polarity mapping A to $B'C'$; B to $A'C'$; C to $A'B'$.

*6. State and prove the converse of Exercise 5.

6. ORDER AND CONTINUITY

Throughout the discussion of order, H will denote an ordered field (Section I.6).

DEFINITION 1 A finite set of points $\{P_1, P_2, \cdots, P_r\}$ on a line l in $\mathbf{P}_nH$ are said to have the *projective order* $\langle P_1, P_2, \cdots, P_r\rangle$ if there exists a frame on l in which $P_1, P_2, \cdots, P_r$ have affine coordinates $p_1, p_2, \cdots, p_r$ satisfying

$$p_{1\pi} < p_{2\pi} < \cdots < p_{r\pi}, \tag{1}$$

for some cyclic permutation π of $(1, 2, \cdots, r)$.

We saw in Section II.1 (Figure II.1.6), that the projective orders $\langle A, B, C, D\rangle$, $\langle B, C, D, A\rangle$, $\langle C, D, A, B\rangle$, $\langle D, A, B, C\rangle$ are apparently indistinguishable. This explains the presence in (1) of the arbitrary cyclic permutation $\pi = (12 \cdots r)^k$, $1 \leq k \leq r$. That Definition 1 actually has the invariance suggested in Figure II.6.1 is a corollary to the following theorem.

THEOREM 1 Collinear points $P_1, P_2, \cdots, P_r$ in $\mathbf{P}_nH$ are in the projective order $\langle P_1, P_2, \cdots, P_r\rangle$ if and only if for every distinct i, j, k between 1 and r, with $j \neq i + 1 \neq k$, the cross ratio

$$(P_iP_{i+1}P_jP_k) > 0, \tag{2}$$

where we take $P_{r+1} = P_1$.

PROOF. Suppose $\langle P_1, P_2, \cdots, P_r\rangle$. Choose a frame in which (1) holds. Then, for $i\pi < r\pi$, $p_{i\pi} - p_{j\pi}$ and $p_{i\pi+1} - p_{j\pi}$ have the same sign, as do $p_{i\pi} - p_{k\pi}$ and $p_{i\pi+1} - p_{k\pi}$. Thus, by Eq. (2.6) $(P_{i\pi}P_{i\pi+1}P_{j\pi}P_{k\pi}) > 0$. Also, $p_{r\pi} - p_{j\pi}$ and $p_{r\pi} - p_{k\pi}$ are positive, $p_{1\pi} - p_{j\pi}$ and $p_{1\pi} - p_{k\pi}$ are negative. Therefore, by Eq. (2.6), $(P_{r\pi}P_{1\pi}P_{j\pi}P_{k\pi}) > 0$. Finally, since π is a cyclic permutation, this establishes (2).

Conversely, suppose (2). Choose a frame in which P_i has affine coordinate p_i. Then

$$p_{1\rho} < p_{2\rho} < \cdots < p_{r\rho} \tag{*}$$

for some permutation ρ of $(1, 2, \cdots, r)$. Let $p_{1\rho} = p_k$. If $p_{k+1} \neq p_{2\rho}$ or $p_{r\rho}$, then by (2.6) $(P_kP_{k+1}P_{2\rho}P_{r\rho}) < 0$, contradicting (2). Therefore, either $p_{k+1} = p_{2\rho}$ or $p_{k+1} = p_{r\rho}$.

ASSERTION. $p_{k+1} \neq p_{r\rho}$. Suppose $p_{k+1} = p_{r\rho}$. If $p_{k-1} \neq p_{2\rho}$, then $(P_{k-1}P_kP_{2\rho}P_{k+1}) < 0$; hence, $p_{k-1} = p_{2\rho}$. If $p_{k-2} \neq p_{3\rho}$, then $(P_{k-2}P_{k-1}P_{3\rho}P_{k+1}) < 0$; hence, $p_{k-2} = p_{3\rho}$. Continuing in this way, we see that (*) must read

$$p_k < p_{k-1} < p_{k-2} < \cdots < p_1 < p_{(k+1)\rho} < \cdots < p_{(r-1)\rho} < p_{k+1}.$$

If $p_{(k+1)\rho} \neq p_r$, then $(P_rP_1P_{(k+1)\rho}P_{k+1}) < 0$; hence, $p_{(k+1)\rho} = p_r$. But then $p_1 < p_r < p_2 < p_{k+1}$, so $(P_1P_2P_rP_{k+1}) < 0$. This contradiction proves the assertion.

So we must have $p_{k+1} = p_{2\rho}$. Now, if $p_{k-1} \neq p_{r\rho}$, then $(P_{k-1}P_kP_{k+1}P_{r\rho}) < 0$; hence, $p_{k-1} = p_{r\rho}$. If $p_{k+2} \neq p_{3\rho}$, then $(P_{k+1}P_{k+2}P_{3\rho}P_{k-1}) < 0$; hence, $p_{k+2} = p_{3\rho}$. Continuing in this way, we see that (*) must read

$$p_k < p_{k+1} < p_{k+2} < \cdots < p_r < p_{(r-k+2)\rho} < \cdots < p_{(r-1)\rho} < p_{k-1}.$$

If $p_1 \neq p_{(r-k+2)\rho}$, then $(P_rP_1P_{(r-k+2)\rho}P_{k-1}) < 0$; hence, $p_1 = p_{(r-k+2)\rho}$. If $p_2 \neq p_{(r-k+3)\rho}$, then $(P_1P_2P_{(r-k+3)\rho}P_{k-1}) < 0$; hence, $p_2 = p_{(r-k+3)\rho}$. Continuing in this way, we see that (*) is

$$p_k < p_{k+1} < \cdots < p_r < p_1 < p_2 < \cdots < p_{k-1},$$

which is (1) for the cyclic permutation $\rho = (123 \cdots r)^k$. Hence, $\langle P_1, P_2, \cdots, P_r\rangle$. ▎

COROLLARY If $\langle P_1, P_2, \cdots, P_r\rangle$ in $\mathbf{P}_nH$ and if μ is a projective collineation of $\mathbf{P}_nH$, then $\langle P_1\mu, P_2\mu, \cdots, P_r\mu\rangle$.

PROOF. Clear from (2), Theorem 4.3, and Definition 4.1. ▎ It is immediate from this that Definition 1 is independent of choice of frame on l.

The projective order of four points is, we have seen, the basis for the general concept. Points A, B are said to *separate* points C, D (see Section II.1) if they are in the order $\langle A, C, B, D\rangle$. The conditions (2) for projective order $\langle A_1, A_2, \cdots, A_r\rangle$ can best be remembered as conditions for $A_1, A_2, \cdots, A_r$ to be in that order (clockwise or counterclockwise) on a circle. The conditions (2) for separation have a particularly short equivalent form.

THEOREM 2 Points A, B separate points C, D in $\mathbf{P}_nH$ if and only if A, B, C, D are collinear and

$$(ABCD) < 0. \tag{3}$$

PROOF. By Theorem 1, $\langle A, C, B, D\rangle$ if and only if $(ACBD) > 0$, $(CBAD) > 0$, and $(ADBC) > 0$. Let $k = (ABCD)$. Then, by the corollary to Theorem 2.4, these conditions become

$$1 - k > 0, \qquad k(k-1)^{-1} > 0, \qquad (k-1)k^{-1} > 0 \tag{*}$$

We must show (*) iff $k < 0$. Suppose $k < 0$. Then $1 - k > 1 > 0$, $k - 1 < 0$, $(k - 1)^{-1} < 0$, $k^{-1} < 0$, and so $k(k - 1)^{-1} > 0$ and $(k - 1)k^{-1} > 0$. Conversely, suppose (*). Then $1 - k > 0 \rightarrow k - 1 < 0 \rightarrow (k - 1)^{-1} < 0$, and so $k(k - 1)^{-1} > 0 \rightarrow k < 0$. ▮

Thanks to (3), properties of the relation of separation can now be read off at will from the corollary to Theorem 2.4. We leave this to the student. The separation relation gives rise to another useful idea.

DEFINITION 2 For A, B, C distinct collinear points of $\mathbf{P}_n H$, the *segments* with *endpoints* A, B are the sets

$$(A, B)_C = \{X \mid A \neq X \neq B \text{ and } A, B \text{ do not separate } C, X\};$$
$$(A, B)_{\sim C} = \{X \mid A, B \text{ separate } C, X\}.$$

Occasional reference is made to the *closed segments* $[A, B]_C$ and $[A, B]_{\sim C}$, obtained by adding A and B to the respective *open* segments defined above. The dependence of the above definition on C is illusory, as we see in the following theorem.

THEOREM 3 Let A, B be distinct points in $\mathbf{P}_n H$. Let C, D be points on AB each distinct from A, B. Then the sets $(A, B)_C$, $(A, B)_{\sim C}$ are the sets $(A, B)_D$, $(A, B)_{\sim D}$ (possibly in opposite order).

PROOF. By Theorem 2, $(A, B)_C = \{X \mid (ABCX) > 0\}$ and $(A, B)_{\sim C} = \{X \mid (ABCX) < 0\}$. By assigning affine coordinates, substituting in Eq. (2.6), and canceling, one obtains the following useful formula:

$$(ABCD)(ABDX) = (ABCX) \tag{*}$$

Now for the proof itself.

CASE 1. $D \in (A, B)_C$. Then $(ABCD) > 0$ and by (*) $(ABDX) > 0$ iff $(ABCX) > 0$. Hence, $(A, B)_C = (A, B)_D$, and, therefore, $(A, B)_{\sim C} = (A, B)_{\sim D}$.

CASE 2. $D \in (A, B)_{\sim C}$. Then $(ABCD) < 0$ and by (*) $(ABDX) > 0$ iff $(ABCX) < 0$. Hence, $(A, B)_{\sim C} = (A, B)_D$, and, therefore, $(A, B)_C = (A, B)_{\sim D}$. ▮

Thus, we may speak simply of "the two segments determined by A, B" on a line.

COROLLARY Under the hypotheses of the theorem, C and D are in different segments determined by A and B if and only if A, B separate C, D.

PROOF. Pick a point E on AB distinct from A, B. By (*)

$$(ABEC)(ABCD) = (ABED).$$

Then $(ABEC)$, $(ABED)$ have opposite signs iff $(ABCD) < 0$, that is, C and D are not both in $(A, B)_E$ or $(A, B)_{\sim E}$ iff A, B separate C, D. ▮

There are many interesting connections between separation and other geometric ideas. We give a proof of one such connection (in n-space) and leave others (in 2-space) as exercises.

THEOREM 4 In $\mathbf{P}_nH$ let $\mathbf{H}$ be a hyperplane and l a line not in $\mathbf{H}$. At least one of the segments determined by any two points on l contains no points of $\mathbf{H}$.

PROOF. Let A, B be two points of l. Let C be the point $l \wedge \mathbf{H}$. If C is A or B, neither segment has a point on $\mathbf{H}$. If $C \neq A, B$, then $(A, B)_{\sim C}$ has no point on $\mathbf{H}$. ▎

* * *

We close this chapter with a brief discussion of convergence and continuity, which will provide useful motivation for Chapter V. Throughout this discussion K will stand for a subfield of the complex numbers. (A less restricted development is possible. The interested reader could begin by studying Chapter X, Vol. I, of [23].) We assume that the reader is familiar with the limit concept of analysis (see [1], Chapter IV, for the real case.)

DEFINITION 3 A sequence of points $\{P_n\}$ in $\mathbf{P}_mK$ *converges* to the point P if there are coordinate vectors $P_n \vdots \mathbf{p}_n$, $P \vdots \mathbf{p}$ such that $\lim_{n\to\infty} \mathbf{p}_n = \mathbf{p}$.

Suppose that another selection of coordinate vectors $P_n \vdots \mathbf{p}'_n$ is made such that $\lim \mathbf{p}'_n = \mathbf{p}' \neq \mathbf{0}$. $\mathbf{p}_n = a_n\mathbf{p}'_n$ for some $a_n \neq 0$. Then $\lim a_n\mathbf{p}'_n$ and $\lim \mathbf{p}'_n$ exist, so $\lim a_n = a$ exists and $\mathbf{p} = \lim a_n\mathbf{p}'_n = a \lim \mathbf{p}'_n = a\mathbf{p}'$. Thus, $P \vdots \mathbf{p}'$, proving that a sequence of points in $\mathbf{P}_mK$ cannot converge to two distinct points. This proves a more general statement of Exercise II.2.6. Next we generalize Exercise II.2.8.

THEOREM 5 Let $\{P_n{}^i\}$, $i = 1, 2, \cdots, m$ be m sequences of points in $\mathbf{P}_mK$ such that for each n the points $P_n{}^1, P_n{}^2, \cdots, P_n{}^m$ are independent. Suppose each $\{P_n{}^i\}$ converges to a point P^i such that $P^1, P^2, \cdots, P^m$ are independent. Then the sequence of hyperplanes $\{[P_n{}^1, P_n{}^2, \cdots, P_n{}^m]\}$ converges, in the sense of the dual of Definition 3, to the hyperplane $[P^1, P^2, \cdots, P^m]$.

PROOF. Choose coordinate vectors so that $\lim \mathbf{p}_n{}^i = \mathbf{p}^i$, $1 \leq i \leq m$. Let $\mathbf{p}_n = (p_{0n}{}^i, p_{1n}{}^i, \cdots, p_{mn}{}^i)$. By Exercise 1.2, $[P_n{}^1, P_n{}^2, \cdots, P_n{}^m]$ has a coordinate vector $\mathbf{u}_n = (u_{0n}, u_{1n}, \cdots, u_{mn})$ where the u_{jn} are determinants involving the $p_{kn}{}^i$. Let $\mathbf{p}^i = (p_0{}^i, p_1{}^i, \cdots, p_m{}^i)$. Then $\lim p_{kn}{}^i = p_k{}^i$ for $0 \leq k \leq m$, $1 \leq i \leq m$. Now a determinant is a continuous function of its entries, so in the limit u_{jn} becomes the same determinant, call it u_j, with the $p_{kn}{}^i$ replaced by $p_k{}^i$. Then, again by Exercise 1.2, $\mathbf{u} = (u_0, u_1, \cdots, u_m)$ is a coordinate vector for $[P^1, P^2, \cdots, P^m]$. Since $\lim u_{jn} = u_j$, $0 \leq j \leq m$, we have $\lim \mathbf{u}_n = \mathbf{u}$, as required by the dual of Definition 3. ▎

The concepts of functional limit and continuity, which will be of use in the next chapter, can now be easily derived from material in analysis.

DEFINITION 4 Let f be a function from $\mathbf{P}_mK$ to itself. Let P, Q be points. Then $\lim_{X\to P} f(X) = Q$ means $\lim_{n\to\infty} f(X_n) = Q$ for all sequences $\{X_n\}$ with limit P; f is *continuous at* P if $\lim_{X\to P} f(X) = f(P)$.

Exercises

1. Prove that three (or fewer) distinct collinear points in $\mathbf{P}_nH$ are in every projective order.
2. Suppose points in $\mathbf{P}_nH$ are in the projective order $\langle P_1, P_2, \cdots, P_r\rangle$. Let π be a permutation of $(1, 2, \cdots, r)$. Prove that $\langle P_{1\pi}, P_{2\pi}, \cdots, P_{r\pi}\rangle$ if and only if π is cyclic.
3. Suppose points in $\mathbf{P}_nH$ are in the projective order $\langle P_1, P_2, \cdots, P_r\rangle$. Prove that $\langle P_i, P_j, P_k, P_l\rangle$ whenever $1 \leq i < j < k < l \leq r$.
4. Let l and m be distinct lines in $\mathbf{P}_2H$. Let A be a point not on l or m. The *sectors* with *boundary lines* l, m are the sets $\mathcal{S}_A = \{B$ not on l or $m \mid A = B$ or A, B do not separate $l \wedge AB, m \wedge AB\}$, and $\mathcal{S}_{\sim A} = \{B$ not on l or $m \mid A, B$ separate $l \wedge AB, m \wedge AB\}$.
State and prove the result for sectors corresponding to Theorem 3.
5. (See Exercise 4.) Prove that two distinct points C, D not on l or m are in the same sector with boundary lines l, m if and only if exactly one of the segments with endpoints C, D contains a point of l or m. (Hence, C, D are in different sectors iff both segments contain a point of l or m.)
6. (See Exercise 4.) If $l \wedge m$ is not on line n, prove that one of the segments with endpoints $l \wedge n$, $m \wedge n$ is in one of the sectors with boundary lines l, m and the other segment is in the other sector.
7. Using Exercise 4, define the *four* (not seven) regions of $\mathbf{P}_2H$ that are determined by three nonconcurrent lines. State and prove results analogous to Exercises 5 and 6.

* * *

8. Generalize Exercise II.2.7 to $\mathbf{P}_mK$.
9. Can the idea of the derivative of a function be given meaning in $\mathbf{P}_mK$?

V

Analytic Geometry

The reader should now study Sections I.7, I.8, and I.9. In the first five sections we shall deal mainly with an arbitrary algebraically closed field K of characteristic zero. The most familiar example of K is the field C of complex numbers.

1. PLANE CURVES

The two methods most often used to describe a curve in the Euclidean plane are the following. (1) Zeros of a function: The curve is $\{(x, y) \mid f(x, y) = 0\}$, where $f\colon R \times R \rightarrow R$. (2) Parametrically: The curve is $\{(x(t), y(t)) \mid t \in R\}$, where $x\colon R \rightarrow R$ and $y\colon R \rightarrow R$. Both of these methods are of use in the analytic geometry of $\mathbf{P}_2K$, but we shall mainly be concerned with (1). Since projective coordinates employ ordered triples, we are interested in solutions of equations of the form $f(x_0, x_1, x_2) = 0$, where $f\colon K \times K \times K \rightarrow K$. In this chapter we restrict our attention to functions f that are polynomials over K. Now, if a point $P\colon(p_0, p_1, p_2)$, then $P\colon(tp_0, tp_1, tp_2)$ for all $t \neq 0$. Therefore, the polynomials f to be studied should have the property: $f(p_0, p_1, p_0) = 0 \rightarrow f(tp_0, tp_1, tp_2) = 0$ for all $t \neq 0$. For this reason, we shall consider only homogeneous polynomials (Section I.7).

DEFINITION 1 Let F be a homogeneous polynomial in three variables over K, with factorization $F = F^1F^2 \cdots F^r$ (Section I.8). The *point curve* $F = 0$ is a list of the sets $\{\text{points } P \text{ in } \mathbf{P}_2K \mid P\colon(p_0, p_1, p_2) \text{ and } F^i(p_0, p_1, p_2) = 0\}$, $i = 1, 2, \cdots, r$. The sets in this list, which by the above are just the curves $F^i = 0$, are called the *prime components* of $F = 0$.

We employ the word "list" instead of "set" in the definition, since repeated components, called *multiple* components, must be counted as often as they occur. If P is on the curve $F = 0$, we write $F(P) = 0$.

The *order* of the point curve $F = 0$ is the degree of the polynomial F. If $F = 0$ and $G = 0$ are the same point curve, then, by Section I.7, $F = kG$ for some constant k. Hence, the order of a point curve in a given coordinate system does not depend on the particular polynomial used to describe it. We go on to show that the concepts defined above are geometrical invariants.

THEOREM 1 Let $F = 0$ be a point curve of order n in $\mathbf{P}_2K$ and let μ be a collineation of $\mathbf{P}_2K$. Then the image of $F = 0$ under μ is a point curve of order n.

PROOF. Suppose that μ has matrix A and automorphism α (Theorem IV.4.4); let $G^i(\mathbf{x}) = F^i((A\alpha^{-1})^{-1}\mathbf{x}\alpha^{-1})$ for each prime factor F^i of F. G^i is a homogeneous prime polynomial of the same degree as F^i, and $F^i(P) = 0$ iff $G^i(P\mu) = 0$. Thus, the image of $F = 0$ is just the nth order curve $G = 0$, where $G = G^1G^2 \cdots G^r$. ▮

Recall the correspondence $(x_0, x_1, x_2) \leftrightarrow (x, y) = (x_0^{-1}x_1, x_0^{-1}x_2)$ between the points of $\mathbf{P}_2K$ not on l_∞:[1, 0, 0] and the points of $\mathbf{A}_2K$. A curve $f(x, y) = 0$ in $\mathbf{A}_2K$, where f is a (not necessarily homogeneous) polynomial, would have the equation $f(x_0^{-1}x_1, x_0^{-1}x_2) = 0$ in the affine part of $\mathbf{P}_2K$. If f is of degree n, then $x_0^n f(x_0^{-1}x_1, x_0^{-1}x_2) = F(x_0, x_1, x_2)$ is a polynomial, and in fact a homogeneous polynomial of degree n. A point (a, b) in $\mathbf{A}_2K$ is on $f = 0$ iff the corresponding point $(1, a, b)$ is on the affine part of $F = 0$. Thus, $F = 0$ is the "projective extension" of $f = 0$. Of course, $F = 0$ may have points on l_∞. We shall illustrate what we have done by three examples.

Line: In $\mathbf{A}_2K$, $f(x, y) = ax + by + c$. In $\mathbf{P}_2K$

$$F(x_0, x_1, x_2) = x_0(ax_0^{-1}x_1 + bx_0^{-1}x_2 + c) = ax_1 + bx_2 + cx_0.$$

Solution on l_∞ is $(0, b, -a)$, as expected from Section II.2.

Parabola: In $\mathbf{A}_2K$, we could, for example, let $f(x, y) = ax^2 + b - y$. Then in $\mathbf{P}_2K$,

$$F(x_0, x_1, x_2) = x_0^2(ax_0^{-2}x_1^2 + b - x_0^{-1}x_2) = ax_1^2 + bx_0^2 - x_0x_2.$$

Solution on l_∞ is $(0, 0, 1)$, where the parabola is "tangent" to l_∞.

Circle: The most general form in $\mathbf{A}_2K$ is $f(x, y) = x^2 + y^2 + ax + by + c$. In $\mathbf{P}_2K$, $F(x_0, x_1, x_2) = x_1^2 + x_2^2 + ax_0x_1 + bx_0x_2 + cx_0^2$. Solutions on l_∞ are $(0, 1, i)$ and $(0, 1, -i)$ where $i, -i$ are the solutions in K of the equation $z^2 + 1 = 0$. Thus, every circle in $\mathbf{A}_2K$ passes through the same two points on l_∞ in $\mathbf{P}_2K$! These points are called the *circular points at infinity.*

Let us try to obtain a good definition of "tangent line to a curve" in $\mathbf{P}_2K$. Suppose P:(p_0, p_1, p_2) is a point on the curve $F = 0$. Changing coordinates if necessary, we may assume $P \notin l_\infty$. Then in $\mathbf{A}_2K$ P:$(a, b) = (p_0^{-1}p_1, p_0^{-1}p_2)$ is on $f(x, y) = 0$, where $x = x_0^{-1}x_1$ and $y = x_0^{-1}x_2$. We want to pick out a tangent line to $f = 0$ from the set of all lines through P. This set of lines may be described as all lines l with parametric equations $x = a + ut$, $y = b + vt$, where each ratio u:v determines a line. For each u:v consider the polynomial $g(t) = f(a + ut, b + vt)$. (Of course, g depends on u:v.) Since P is on $f = 0$, $g(0) = 0$ for all u:v. The line that "sticks the tightest" to $f = 0$ at P would be the line for which $g(t)$ has $t = 0$ as a root of multiplicity greater than one. Is there necessarily such a line and is it unique? By the chain rule (Section I.7) $g'(0) = uf_x(a, b) + vf_y(a, b)$. Thus, $g'(0) = 0$ will always have a solution $u:v$, and it will be unique unless $f_x(a, b) = f_y(a, b) = 0$. Let us translate these last equations back into projective terms. We have $f_x(a, b) = F_1(1, a, b)$ and $f_y(a, b) = F_2(1, a, b)$. By Euler's theorem (Section I.8), $F_0(1, a, b) + aF_1(1, a, b) + bF_2(1, a, b) = nF(1, a, b) = 0$, where n is the degree of F. Hence, $f_x = f_y = 0$ at P iff $F_0 = F_1 = F_2 = 0$ at P.

DEFINITION 2 A point P of a point curve $F = 0$ is *singular* if $F_0(P) = F_1(P) = F_2(P) = 0$. The curve is called singular if it has at least one singular point.

With the above notation suppose P is a nonsingular point of $F = 0$. Then the tangent line at P is the unique line $x = a + ut$, $y = b + vt$ for which $uF_1(1, a, b) + vF_2(1, a, b) = 0$. Multiplying $x - a$ by $F_1(P)$, $y - b$ by $F_2(P)$, adding, and using the three previous equations, we obtain the equation for the line in $\mathbf{A}_2K$: $F_1(P)(x - a) + F_2(P)(y - b) = 0$. But, as we saw before, $F_0(P) + aF_1(P) + bF_2(P) = 0$. Thus, the tangent line has the equation $F_1(P)x + F_2(P)y + F_0(P) = 0$. In projective coordinates, then, we have another definition.

DEFINITION 3 The *tangent* to a point curve $F = 0$ at a nonsingular point P is the line $[F_0(P), F_1(P), F_2(P)]$.

Consider, for example, curve $F = 0$ where $F(x_0, x_1, x_2) = x_1^2 - x_1x_2$. We have $F_0 = 0$, $F_1 = 2x_1 - x_2$, and $F_2 = -x_1$. The only singular point is $(1, 0, 0)$. The tangent at the nonsingular point $(1, 0, 1)$ is $[0, 1, 0]$; the tangent at $(1, 1, 1)$ is $[0, 1, -1]$. Closer examination of F reveals the curve to consist of the two lines $[0, 1, 0]$ and $[0, 1, -1]$. The singular point $(1, 0, 0)$ is just the intersection of these lines.

The proof of geometric invariance for Definitions 2 and 3 are left as exercises, as is the dualization of the above.

Exercises

1. Write out the duals of the definitions of this section. The duals of "point curve" and "tangent" are called, respectively, "line curve" and "point of contact."
2. Verify the "tangent" statement in the *parabola* example of this section.
3. Investigate the converse of the result obtained in the *circle* example of this section.
4. Prove that the point curve $x_0^2x_1 + kx_2^3 + x_0x_1x_2 = 0$ has one singular point if $k \neq 0$ and three singular points if $k = 0$.

5. (See 1.) Show that the set of all tangents to the nonsingular point curve $x_0^2 + x_1x_2 = 0$ is just the nonsingular line curve $u_0^2 + 4u_1u_2 = 0$, and that the set of all points of contact of this line curve is just the original point curve.
*6. (See 5.) Does the set of all tangents to a nonsingular point curve necessarily form a line curve?
7. Prove that the image under a collineation of $\mathbf{P}_2K$ of a singular (nonsingular) point of a point curve is a singular (nonsingular) point of the image curve.
8. (See 7.) Prove that the image of the tangent to a point curve at a point is the tangent to the image curve at the image point.
9. A *component* of a curve is any curve that is contained in the given curve. (Only "prime component" has previously been defined.) Prove the geometrical invariance of this concept.

2. CONICS

A *conic* is a plane curve of order two. Suppose $F(x_0, x_1, x_2) = \sum_{i,j=0}^{2} b_{ij}x_ix_j = 0$ is a conic. Let $a_{ij} = \frac{1}{2}(b_{ij} + b_{ji})$. Then $F = \sum a_{ij}x_ix_j$ and $a_{ij} = a_{ji}$. We shall always assume F to be written in this symmetric way. The polynomial F is a quadratic form. Although many results on conics are easy consequences of the theory of quadratic forms, we shall proceed without reference to this theory.

The only possible factorization of a polynomial of degree two is into two linear polynomials. This leads to the following theorem.

THEOREM 1 If a conic has a proper component, then the conic consists of either one or two lines. If it consists of one line, then every point is singular. If it consists of two lines, then only their point of intersection is singular, and the tangent at any other point is the line containing it.

PROOF. Exercise 1.

The conic $\sum a_{ij}x_ix_j = 0$ can be expressed neatly by means of the 3×3 matrix $A = (a_{ij})$ and vector $\mathbf{x} = (x_0, x_1, x_2)$. The equation then becomes $\mathbf{x}\,A\,\mathbf{x} = 0$. Since $a_{ij} = a_{ji}$, A is a *symmetric* matrix, that is, $A = A^T$. A is determined by the conic up to a nonzero scalar multiple. Much of the theory of conics is best expressed in this form.

THEOREM 2 Let $\mathbf{x}\,A\,\mathbf{x} = 0$ be a point conic. Then: (1) if $P{:}\mathbf{p}$ is a nonsingular point, the tangent at P has coordinate $A\mathbf{p}$; (2) the conic is singular if and only if A is singular; (3) if the conic is nonsingular, then the set of all its tangents forms the line conic $\mathbf{l}\,A^{-1}\,\mathbf{l} = 0$.

PROOF

(1) If $F = \mathbf{x}\,A\,\mathbf{x} = \sum a_{ij}x_ix_j$, then $F_i = 2\sum_j a_{ji}x_j$, $i = 0, 1, 2$. Thus $[F_0, F_1, F_2] = 2\,A\,\mathbf{x}$. (The 2 is irrelevant.)

(2) Recall that A is singular iff there exists a vector $\mathbf{x} \neq \mathbf{0}$ such that $A\mathbf{x} = \mathbf{0}$. The result now follows from the last equation in the proof of (1).

(3) If l:$\mathbf{l}$ is tangent at P:$\mathbf{p}$, then $\mathbf{l} = t\,A\,\mathbf{p}$ for some $t \neq 0$. Then $\mathbf{l}\,A^{-1}\,\mathbf{l} = t^2\mathbf{p}\,AA^{-1}\,A\mathbf{p} = t^2\mathbf{p}\,A\,\mathbf{p} = 0$. Conversely, if $\mathbf{l}\,A^{-1}\,\mathbf{l} = 0$, then P:$\mathbf{p} = \mathbf{l}\,A^{-1}$ is on the conic $\mathbf{x}\,A\,\mathbf{x} = 0$, and the tangent at P has coordinate $A\mathbf{p} = \mathbf{l}$. ∎

A collineation with matrix M and automorphism α maps the conic $\mathbf{x}\,A\,\mathbf{x} = 0$ to the conic $\mathbf{x}(M^{T-1}\,A\,\alpha M^{-1})\mathbf{x} = 0$ (Exercise 3). By a proper choice of M and α, the equation for the conic could be simplified. We shall accomplish this in a rather more geometric way.

DEFINITION 1 Points P:$\mathbf{p}$ and Q:$\mathbf{q}$ are *conjugate* with respect to the conic $\mathbf{x}\,A\,\mathbf{x} = 0$ if $\mathbf{p}\,A\,\mathbf{q} = 0$.

The self-conjugate points are precisely the points of the conic itself. The set of all points conjugate to a given point P is called the *polar* of P. (The dual notion is *pole*.) The polar of a singular point consists of the entire plane. The polar of any other point, that is, of a nonsingular point or a point not on the conic, is a line, as we shall see in the proof of Theorem 4. The geometric interpretation of conjugacy makes use of Definition IV.3.3.

THEOREM 3 Let A and B be distinct points on a conic. A and B are conjugate if and only if AB is a component. If AB is not a component, then points C, D on AB are conjugate if and only if $H(AB, CD)$.

PROOF. Choose coordinates so that $\mathbf{a} + \mathbf{b} = \mathbf{c}$ and $\mathbf{a} + \lambda\mathbf{b} = \mathbf{d}$. Since A and B are on the conic $\mathbf{x}\,A\,\mathbf{x} = 0$, we have $\mathbf{c}\,A\,\mathbf{d} = (1 + \lambda)\mathbf{a}\,A\,\mathbf{b}$. If A and B are conjugate, then, taking $C = D$, we have $\mathbf{c}\,A\,\mathbf{c} = 0$, that is, C is on the conic, for all C on AB. Thus, AB is a component. Conversely, if AB is a component, then for any C on AB, $\mathbf{c}\,A\,\mathbf{c} = 2\,\mathbf{a}\,A\,\mathbf{b} = 0$. Thus, $\mathbf{a}\,A\,\mathbf{b} = 0$ and A, B are conjugate. If A and B are not conjugate, then $\mathbf{c}\,A\,\mathbf{d} = 0$ iff $\lambda = -1$, that is, iff $H(AB, CD)$ (Theorem IV.3.2). ∎

A set of points is *self-polar* with respect to a conic if any two distinct points in the set are conjugate with respect to the conic.

THEOREM 4 Given any conic, there is a simplex (Definition IV.2.1) that is self-polar with respect to it.

PROOF. Choose a point Q_0:$\mathbf{q}_0$ not on the conic $\mathbf{x}\,A\,\mathbf{x} = 0$. $A\mathbf{q}_0 \neq \mathbf{0}$; therefore, the vector space $\{\mathbf{x} \mid \mathbf{x}\,A\,\mathbf{q}_0 = 0\}$ has dimension two. Hence, the polar of Q_0 is a line l_0 that does not contain Q_0.

CASE 1. l_0 is a component of the conic. Choose any two distinct points Q_1, Q_2 on l_0. $\{Q_0, Q_1, Q_2\}$ is a self-polar simplex.

CASE 2. l_0 is not a component. Choose a point Q_1 on l_0 not on the conic. As above, the polar of Q_1 is a line l_1 not through Q_1. Let $Q_2 = l_0 \wedge l_1$. $\{Q_0, Q_1, Q_2\}$ is a self-polar simplex. ▎

With this we can obtain simple forms for the equations of conics.

THEOREM 5 Given a conic, there exists a coordinate system in which it has the equation: (1) $x_0^2 + x_1^2 + x_2^2 = 0$, if it is nonsingular; (2) $x_0^2 + x_1^2 = 0$, if it has distinct components; (3) $x_0^2 = 0$, if it has a multiple component. There are other coordinate systems in which the conics in Cases (1) and (2) have equations $x_0^2 - x_1x_2 = 0$, $x_0x_1 = 0$, respectively.

PROOF. Let the conic be $\mathbf{x}\, A\, \mathbf{x} = \sum a_{ij}x_ix_j = 0$. Let $\{Q_0, Q_1, Q_2\}$ be a self-polar simplex. In any frame of reference with this simplex as coordinate simplex $Q_0\!:\!\mathbf{e}_0 = (1, 0, 0)$, $Q_1\!:\!\mathbf{e}_1 = (0, 1, 0)$, $Q_2\!:\!\mathbf{e}_2 = (0, 0, 1)$. Since $\mathbf{e}_i\, A\, \mathbf{e}_j = 0$ for $i \neq j$, we have $a_{ij} = 0$ for $i \neq j$. Thus, the equation for the conic in such a frame is $\sum a_{ii}x_i^2 = 0$. If the conic is nonsingular, $a_{ii} \neq 0$ for $i = 0, 1, 2$. Keeping Q_0, Q_1, Q_2, choose a unit point so that a point with old coordinates (x_0, x_1, x_2) has new coordinates $(a_{00}^{1/2}x_0, a_{11}^{1/2}x_1, a_{22}^{1/2}x_2)$. In this coordinate system the equation is (1). If the conic is singular, then in any frame with simplex $\{Q_0, Q_1, Q_2\}$ at least one a_{ii} is zero, but not all a_{ii} are zero. Relabeling the Q_i if necessary and then choosing a unit point as above, we obtain (2) or (3). Since the polynomial in Eq. (1) has no linear factors, the one in (2) has distinct factors, and the one in (3) has repeated factors, these equations correspond to the types of conics mentioned. The alternate forms of (1) or (2) are obtained by coordinate changes in which x_1 is replaced by $\frac{1}{2}(x_1 - x_2)$, x_2 by $(i/2)(x_1 + x_2)$, or x_0 by $\frac{1}{2}(x_0 + x_1)$, x_1 by $(i/2)(x_0 - x_1)$, respectively, where i is a root of the equation $z^2 + 1 = 0$. ▎

We then have two corollaries.

COROLLARIES

1. A conic is singular if and only if it has a component. Indeed, a conic $\mathbf{x}\, A\, \mathbf{x} = 0$ can be given Eq. (1), (2), or (3) above if and only if rank A is 3, 2, or 1, respectively.
2. All nonsingular point conics in $\mathbf{P}_2K$ are projectively equivalent.

The next theorem will be of help in obtaining two classical characterizations of nonsingular conics.

THEOREM 6 The tangent at a point P of a nonsingular conic does not again intersect the conic, but every other line through P intersects the conic at exactly one other point.

PROOF. Let $\sum a_{ij}x_ix_j = 0$ be the conic, let $Q \neq P$ be on the tangent at P, and let R be a point of the conic not on PQ. Choose coordinates in which P:(0, 1, 0), Q:(0, 0, 1), R:(1, 0, 0). In this coordinate system $a_{00} = a_{11} = 0$, $a_{22} \neq 0$, and the tangent at P has coordinate [1, 0, 0]. The equation of the conic is then of the form $x_0x_1 + ax_0x_2 + bx_2^2 = 0$, where $b \neq 0$. If $x_0 = 0$, this equation implies $x_2 = 0$; hence, the tangent meets the conic only at P. In the associated affine coordinates the conic has the equation $x + ay + by^2 = 0$, and an ordinary line through P has the equation $y = c$. These two equations have a unique common solution. ∎

The following result is due in part to J. Steiner (1832). We denote by $(ABCD)_X$ the cross ratio of the lines AX, BX, CX, DX.

THEOREM 7 Let A, B, C, D be four points, no three collinear, and let k be a constant different from 0 and 1. Then

$$\{A, B, C, D\} \cup \{X \mid (ABCD)_X = k\} \tag{1}$$

is a nonsingular point conic. Conversely, suppose A, B, C, D, E are distinct points on a nonsingular conic. Let $k = (ABCD)_E$. Then k is different from 0 and 1, and (1) is the given conic.

PROOF. By Theorem 6 we may assume in either half of the proof that A:(1, 0, 0), B:(0, 1, 0), C:(0, 0, 1), D:(1, 1, 1). If X:$(x_0x_1x_2)$, then AX:$\mathbf{a} = [0, x_2, -x_1]$, BX:$\mathbf{b} = [x_2, 0, -x_0]$, CX:$\mathbf{c} = [x_1, -x_0, 0]$, DX:$\mathbf{d} = [x_1 - x_2, x_2 - x_0, x_0 - x_1]$. Now $-x_0\mathbf{a} + x_1\mathbf{b} = x_2\mathbf{c}$ and $(x_2 - x_0)\mathbf{a} + (x_1 - x_2)\mathbf{b} = x_2\mathbf{d}$. Hence,

$$(ABCD)_X = \frac{x_1(x_0 - x_2)}{x_0(x_1 - x_2)}. \tag{*}$$

The equation $(ABCD)_X = k$ becomes $x_1(x_0 - x_2) - kx_0(x_1 - x_2) = 0$, which is in fact a conic through A, B, C, D. The determinant of the matrix of this form is $\frac{1}{4}k(k - 1)$; hence, by Theorem 2(2) the conic is singular iff k is 0 or 1. Conversely, suppose that the conic $\sum a_{ij}x_ix_j = 0$ passes through A, B, C, D, and E. Then $a_{ii} = 0$ for all i, and $a_{01} + a_{02} + a_{12} = 0$. Since the conic is nonsingular, a_{01}, a_{02}, and a_{12} are nonzero. An equation of the conic is, therefore, $a_{12}x_1(x_0 - x_2) + a_{02}x_0(x_1 - x_2) = 0$. By (*) then $(ABCD)_X = -a_{02}/a_{12}$ for any X on the conic different from A, B, C, D. Setting $X = E$, we have $-a_{02}/a_{12} = k$. If $k = 0$, then $a_{02} = 0$; if $k = 1$, then $a_{01} = 0$. Hence, $0 \neq k \neq 1$. Then X is on the conic iff $(ABCD)_X = k$. ∎

Additional result: In the second part of the theorem, $(ABCD)_X = k$ even if $X = A$, B, C, or D, where "line PP" is interpreted as the tangent at P.

PROOF. As above, we may assume the conic is $a_{12}x_1(x_0 - x_2) + a_{02}x_0(x_1 - x_2) = 0$. Say $X = A$:(1, 0, 0). Then AA:$[0, a_{12} + a_{02}, -a_{02}]$, AB:[0, 0, 1], AC:[0, 1, 0], AD:[0, 1, −1], and $(ABCD)_A = -\dfrac{a_{02}}{a_{12}} = k$. ∎

The following theorem, in the dual formulation, is partly due to M. Chasles (1828).

THEOREM 8 Let A and B be distinct points on a nonsingular conic. Denote by t_A, t_B the tangents at A, B, respectively. Define $\mu: \tilde{A} \to \tilde{B}$ as follows: $\mu(t_A) = AB$, $\mu(AB) = t_B$ and for any other line l of $\tilde{A}$ $\mu(l) = BL$ where L is the other point where l meets the conic (Theorem 6). Then μ is a projectivity. Conversely, let A and B be distinct points and $\mu: \tilde{A} \to \tilde{B}$ a projectivity that does not fix AB. Then

$$\{l \wedge \mu(l) \mid l \in \tilde{A}\} \tag{2}$$

is a nonsingular point conic with $\mu^{-1}(AB)$ the tangent at A and $\mu(AB)$ the tangent at B.

PROOF. By Theorem 6 μ is well-defined, one–one, and onto. By Theorem 7 and the additional result, μ preserves cross ratio. Hence, μ is a projectivity. Conversely, suppose that $\mu: \tilde{A} \to \tilde{B}$ is a projectivity that does not fix AB. Let $t_A = \mu^{-1}(AB)$, $t_B = \mu(AB)$. Let l be a line through A different from t_A and AB. A point X is in the set (2) if and only if

$$(AB\, t_A\, l\, AX) = (t_B\, AB\, \mu(l)\, BX). \tag{*}$$

We may assume $A⁝(0, 0, 1)$, $AB⁝[1, 0, 0]$, $t_A⁝[0, 1, 0]$, $t_B⁝[0, 0, 1]$ and $l⁝[1, 1, 0]$. If $X⁝(x_0, x_1, x_2)$, then $AX⁝[x_1, -x_0, 0]$, and so $(ABt_AlAX) = -x_1/x_0$. Now, $\mu(t_A) = AB$, $\mu(AB) = t_B$ and $l⁝[0, 1, 0] + [1, 0, 0]$. Therefore, $\mu(l)⁝[1, 0, 0] + [0, 0, 1] = [1, 0, 1]$. Also, $B⁝(0, 1, 0)$, and so $BX⁝[x_2, 0, -x_0]$. Thus, $(t_BAB\mu(l)BX) = -x_0/x_2$. By (*), then $-x/x_0 = -x_0/x_2$, that is, $x_0^2 - x_1x_2 = 0$. Hence, (2) is a nonsingular conic. ▮

One further result concerning conics is obtained in the corollary to Theorem 4.3. For more results see [14], [22], [19].

Exercises

1. Prove Theorem 1. In the first part of the proof show that in fact every point on a multiple component of a curve of *any* degree is singular.
2. In contrast to Theorem 1(3), give an example of a nonsingular point curve of degree three the tangents of which form a nonsingular line curve of degree greater than three. (This latter degree is called the *class* of the point curve.)
3. Verify the statement made just after the proof of Theorem 2.
4. Prove that if a line contains three points of a conic it is a component.
5. Prove that every line and every conic in $\mathbf{P}_2K$ intersect in two points (properly counted).

6. Prove that every conic in $\mathbf{P}_2\mathrm{R}$ can be given one of the following equations: (1) $x_0^2 + x_1^2 + x_2^2 = 0$ (the empty set); $x_0^2 + x_1^2 - x_2^2 = 0$ (real nonsingular); $x_0^2 - x_1^2 = 0$ (two distinct lines); $x_0^2 + x_1^2 = 0$ (one point); $x_0^2 = 0$ (one line counted twice.)

3. QUADRICS

A *quadric** is the n-dimensional version of a point conic. It consists of the points in $\mathbf{P}_nK$ satisfying a second-degree homogeneous polynomial in $n + 1$ variables over K, with components handled as for conics. The elementary theory of quadrics is essentially the same as for conics, although some proofs become more complicated. We present here a part of this theory as a simple example of n-dimensional analytic geometry.

It is often convenient to write the equation of a quadric in the form $\mathbf{x}\,A\,\mathbf{x} = 0$, where $\mathbf{x} = (x_0, x_1, \cdots, x_n)$ and A is an $(n+1) \times (n+1)$ symmetric matrix.

A line l is *tangent* to a quadric at a point P, on l and the quadric, if l either is contained in the quadric or intersects it only at P. The *tangent space* at P consists of P together with all points $Q \neq P$ such that QP is tangent to the quadric at P.

Let $P{:}\mathbf{p}$ be a point on the quadric $\mathbf{x}\,A\,\mathbf{x} = 0$. Let $Q{:}\mathbf{q}$ be a point distinct from P. Each point $X \neq P$ on PQ has a coordinate of the form $x\mathbf{p} + \mathbf{q}$. Then X is on $\mathbf{x}\,A\,\mathbf{x} = 0$ iff $2x\mathbf{q}\,A\,\mathbf{p} + \mathbf{q}\,A\,\mathbf{q} = 0$. By definition, then, Q belongs to the tangent space iff $2x\mathbf{q}\,A\,\mathbf{p} + \mathbf{q}\,A\,\mathbf{q}$ is either zero for all x or never zero. This occurs iff $\mathbf{q}\,A\,\mathbf{p} = 0$ (why?). Thus, the tangent space at P consists of all Q such that $\mathbf{q}\,A\,\mathbf{p} = 0$. (Note that this includes $Q = P$.) If $A\mathbf{p} = \mathbf{0}$ the tangent space at P would be all of $\mathbf{P}_nK$; in this case P is *singular*. Otherwise the tangent space at P is the hyperplane with coordinate $A\mathbf{p}$, called the *tangent hyperplane* at P. The quadric is called singular if it has at least one singular point. We have extended Theorem 2.2 to quadrics, and have obtained the following theorem.

THEOREM 1 Let $\mathbf{x}\,A\,\mathbf{x} = 0$ be a quadric. Then: (1) if $P{:}\mathbf{p}$ is a nonsingular point, the tangent hyperplane at P has coordinate $A\mathbf{p}$; (2) the quadric is singular if and only if A is singular; (3) if the quadric is nonsingular, then the set of all its tangent hyperplanes forms the dual quadric (called a *hypersurface* of *class two*) $\mathbf{u}\,A^{-1}\,\mathbf{u} = 0$.

Two points $P{:}\mathbf{p}$, $Q{:}\mathbf{q}$ are *conjugate* with respect to the quadric $\mathbf{x}\,A\,\mathbf{x} = 0$ if $\mathbf{p}\,A\,\mathbf{q} = 0$.

---

* Some authors say *hyperquadric*. Do not confuse quadric with *quartic* (a curve of degree four).

THEOREM 2 Let A and B be distinct points on a quadric. A and B are conjugate if and only if AB is contained in the quadric. Otherwise, points C, D on AB are conjugate if and only if $H(AB, CD)$.

PROOF. See proof of Theorem 2.3.

Let P be a point not on a given quadric. The set of all points conjugate to P forms a hyperplane called the *polar* of P. The polar of a nonsingular point on a quadric is just the tangent hyperplane. A set of points is *self-polar* with respect to a given quadric if any two distinct points of the set are conjugate.

THEOREM 3 Given any quadric, there is a simplex that is self-polar with respect to it.

Before the proof we state a lemma.

LEMMA Suppose $\mathbf{L}$, $\mathbf{M}$, and $\mathbf{N}$ are subspaces of $\mathbf{P}_nK$ such that $\mathbf{L} \wedge \mathbf{M} = \emptyset$ and $\mathbf{N} \subset \mathbf{M}$. Let P be a point in $\mathbf{M}$ that is not in $\mathbf{N}$. Then $P\mathbf{L} \wedge \mathbf{N} = \emptyset$.

PROOF. Exercise 3.

PROOF OF THEOREM. Let $\mathfrak{K}$ be a quadric. Consider the following statement. $\mathfrak{A}(k)$: there exist k independent points $Q_0, Q_1, \cdots, Q_{k-1}$ and an $(n-k)$-dimensional subspace $\mathbf{L}_{n-k}$ such that

(1) $\{Q_0, Q_1, \cdots, Q_{k-1}\}$ is self-polar.
(2) Every Q_i is conjugate to every point in $\mathbf{L}_{n-k}$.
(3) $[Q_0, Q_1, \cdots, Q_{k-1}] \wedge \mathbf{L}_{n-k} = \emptyset$.

We prove $\mathfrak{A}(k)$ for $1 \leq k \leq n+1$ by induction on k. For $\mathfrak{A}(1)$ pick any point Q_0 not in $\mathfrak{K}$ and let $\mathbf{L}_{n-1}$ be its polar. Items (1), (2), and (3) are then clear. Now, suppose we have $\mathfrak{A}(k-1)$ for points $Q_0, Q_1, \cdots, Q_{k-2}$ and subspace $\mathbf{L}_{n-k+1}$. We must find Q_{k-1} and $\mathbf{L}_{n-k}$ for $\mathfrak{A}(k)$.

CASE 1. $\mathbf{L}_{n-k+1} \subset \mathfrak{K}$. If A and B are distinct points of $\mathbf{L}_{n-k+1}$, then AB is contained in $\mathfrak{K}$ and hence by Theorem 2 is self-polar. Let $R_0, R_1, \cdots, R_{n-k+1}$ be a basis of $\mathbf{L}_{n-k+1}$. Then $Q_0, Q_1, \cdots, Q_{k-2}, R_0, R_1, \cdots, R_{n-k+1}$ are independent and therefore form a basis of $\mathbf{P}_nK$. Every point in $\mathbf{L}_{n-k+1}$ is conjugate to the R_i and, by (2), to the Q_i. Hence, each point in $\mathbf{L}_{n-k+1}$ is conjugate to all points of $\mathbf{P}_nK$, that is, every point of $\mathbf{L}_{n-k+1}$ is singular. Now choose any point Q_{k-1} in $\mathbf{L}_{n-k+1}$ and any $(n-k)$-dimensional subspace $\mathbf{L}_{n-k}$ of $\mathbf{L}_{n-k+1}$ that does not contain Q_{k-1}. Then (1) and (2) of $\mathfrak{A}(k)$ are obvious, and (3) follows from the lemma.

CASE 2. $\mathbf{L}_{n-k+1} \not\subset \mathfrak{K}$. Choose Q_{k-1} in $\mathbf{L}_{n-k+1}$ not on $\mathfrak{K}$. Let $\mathbf{H}$ be the polar of Q_{k-1}. Since Q_{k-1} is not on $\mathfrak{K}$, it is not on $\mathbf{H}$. Therefore, $\mathbf{L}_{n-k+1}$ is not a subspace of $\mathbf{H}$, and, hence, $\mathbf{L}_{n-k} = \mathbf{H} \wedge \mathbf{L}_{n-k+1}$ is of dimension $n-k$. Again, (1) and (2) of $\mathfrak{A}(k)$ are obvious and (3) follows from the lemma.

We have established $\mathfrak{A}(k)$ for $1 \leq k \leq n + 1$. Statement (1) of $\mathfrak{A}(n + 1)$ is the theorem. ∎

Let $\sum a_{ij}x_ix_j = 0$ be a quadric and $(Q_0Q_1 \cdots Q_n|U)$ a frame whose simplex is self-polar with respect to it. In this frame $a_{ij} = 0$ for all $i \neq j$ and, reordering the simplex if required, the equation for the quadric becomes $\sum_{i=0}^{r-1} b_ix_i^2 = 0$, with all b_i nonzero. r is the rank of the matrix (a_{ij}) and is also called the *rank of the quadric*. Finally, by proper selection of U, the equation may be simplified to $\sum_{i=0}^{r-1} x_i^2 = 0$. There are, therefore, exactly $n + 1$ projectively nonequivalent quadrics in $\mathbf{P}_nK$.

Exercises

1. State and prove the analog to Theorem 2.1 for quadrics.
2. Given a quadric and a point P, which may or may not be on it, prove that the intersection of the set of all polars of points in the polar of P is just P itself.
3. Prove the lemma to Theorem 3.
4. Show that a given quadric in $\mathbf{P}_n\mathrm{R}$ has in some coordinate system an equation of the form $\sum_{i=1}^{k} x_i^2 = \sum_{j=k+1}^{k+t} x_j^2$, where $1 \leq k$, $0 \leq t \leq k$, and $k + t \leq n + 1$.
5. Show that the number k of Exercise 4 has the following geometrical significance: (a) there is a subspace of $\mathbf{P}_n\mathrm{R}$ of dimension k, but none of higher dimension, which is disjoint from the quadric; (b) there is a subspace of $\mathbf{P}_n\mathrm{R}$ of dimension $n - k - 1$, but none of higher dimension, which is contained in the quadric.
6. Show that the number of projectively nonequivalent quadrics in $\mathbf{P}_n\mathrm{R}$ is $\frac{1}{4}(n + 2)(n + 4)$ if n is even and $\frac{1}{4}(n + 3)^2$ if n is odd.
7. Define the *pole* of a hyperplane with respect to a quadric. Given a nonsingular quadric, consider the function κ defined on points by $P\kappa$ = polar of P, and on hyperplanes by $\mathbf{H}\kappa$ = pole of $\mathbf{H}$. Is κ a correlation? an involution? a polarity?

4. SINGULARITIES AND INTERSECTIONS OF PLANE CURVES

We now continue the discussion of singular points on plane curves, which started in Section 1. Let P be on the point curve $F = 0$. In appropriate affine coordinates, P:(a, b) and the curve has equation $f(x, y) = 0$. We seek those lines l: $x = a + ut$, $y = b + vt$ for which $g(t) = f(a + ut, b + vt)$ has a multiple root at $t = 0$.

There exists an integer $k \geq 1$ such that all derivatives of f of order less than k are zero at (a, b), whereas at least one kth-order derivative of f is not zero at (a, b). From Section I.7,

$$g^{(s)}(0) = u^s f_{x^s}(a, b) + \binom{s}{1} u^{s-1} v f_{x^{s-1}y}(a, b) + \cdots + v^s f_{y^s}(a, b).$$

Thus, for every (u, v), g has at least a $(k - 1)$-fold root at $t = 0$. The equation $g^{(k)}(0) = 0$ has, by Section I.8, exactly k roots $u{:}v$, counting repetitions. Geometrically speaking, then, every line through P has at P a $(k - 1)$-fold intersection with the curve, except for k lines (properly counted) that have k-fold intersections. P is then called a point of *multiplicity* k, and the k special lines are the *tangents* at P. The interpretation of multiplicity k in homogeneous coordinates is an extension of the result obtained in Section 1 for $k = 1$.

THEOREM 1 A point P of a curve $F = 0$ is of multiplicity k if and only if all derivatives of F of order less than k, but not all derivatives of order k, vanish at P.

PROOF. Exercise 1.

If two lines have two points in common they are identical; if a line and a conic have three points in common they have a common component. These observations are special cases of the following general theorem of E. Bezout (1730–1783).

THEOREM 2 If two curves of orders m and n have more than mn distinct points in common, they have a common component.

PROOF. Suppose the curves have more than mn common points. Select $mn + 1$ of these points and choose coordinates so that the point with coordinate $(0, 0, 1)$ is not collinear with any pair of them and is on neither curve. In these coordinates the curves have equations of the form $F(\mathbf{x}) = x_2{}^m + A_1(x_0, x_1)x_2{}^{m-1} + \cdots + A_m(x_0, x_1) = 0$, $G(\mathbf{x}) = x_2{}^n + B_1(x_0, x_1)x_2{}^{n-1} + \cdots + B_n(x_0, x_1) = 0$, where A_i, B_j are either zero or homogeneous of degree i, j, respectively. Let $R(x_0, x_1)$ be the resultant of F and G (Section I.7). Consider

$$R(tx_0, tx_1) = \begin{vmatrix} t^m A_m & t^{m-1}A_{m-1} & \cdots 1 & \cdots & 0 \\ 0 & t^m A_m & \cdots & & \\ \vdots & & & & \\ t^n B_n & \cdots & 1 & \cdots & 0 \\ \vdots & & & & \\ 0 & \cdots & t^n B_n & \cdots & 1 \end{vmatrix}$$

Multiply the ith row of A's by t^{n-i+1} and the jth row of B's by t^{m-j+1} for $1 \leq i \leq n$, $1 \leq j \leq m$. Then the kth column has a factor of $t^{m+n-k+1}$, $1 \leq k \leq m + n$, which we extract. We have multiplied $R(tX_0, tX_1)$ by t^p, where

$p = \sum_{i=1}^{n} (n - i + 1) + \sum_{j=1}^{m} (m - j + 1) = \frac{1}{2}[n(n + 1) + m(m + 1)]$, and have extracted a factor of t^q, where $q = \sum_{k=1}^{m+n} (m + n - k + 1) = \frac{1}{2}(m + n)(m + n + 1)$, leaving just $R(x_0, x_1)$. In other words, we have $t^p R(tx_0, tx_1) = t^q R(x_0, x_1)$. But $q - p = mn$; hence,

$$R(tx_0, tx_1) = t^{mn} R(x_0, x_1). \qquad (*)$$

Now let (a_0, a_1, a_2) be the coordinate of one of the $mn + 1$ common points selected above. The polynomials $f(x_2) = x_2^m + A_1(a_0, a_1)x_2^{m-1} + \cdots + A_m(a_0, a_1)$ and $g(x_2) = x_2^n + B_1(a_0, a_1)x_2^{n-1} + \cdots + B_n(a_0, a_1)$ have the common root a_2, so by Section I.7 their resultant, which is just $R(a_0, a_1)$, is zero. Let (b_0, b_1, b_2) be the coordinate of another of the $mn + 1$ common points. If $a_0{:}a_1 = b_0{:}b_1$, we would have (a_0, a_1, a_2), (b_0, b_1, b_2), and $(0, 0, 1)$ collinear, contradicting our choice of $(0, 0, 1)$. Therefore, we have $R(x_0, x_1) = 0$ for $mn + 1$ distinct ratios. But by (*), $R(x_0, x_1)$ is either homogeneous of degree mn or identically zero. Hence, by Section I.8, $R(x_0, x_1)$ is identically zero, and by Section I.7, F and G have a common factor. ▌

We now consider the critical case in which two curves have the greatest possible number of common points without having a common component.

THEOREM 3 If two curves of order n intersect in n^2 distinct points and if exactly mn of these lie on a prime curve of order m, then the remaining $n^2 - mn$ points lie on a curve of order $n - m$.

PROOF. Let $F = 0$ and $G = 0$ be the curves of order n and $H = 0$ be the prime curve of order m. Pick a point P on $H = 0$ not on $F = 0$ or $G = 0$. Choose nonzero constants a and b such that $aF(P) + bG(P) = 0$. Then the curves $aF + bG = 0$ and $H = 0$ have at least $mn + 1$ common points, and, hence, by Bezout's theorem they have a common component. Since H is prime, we must therefore have that H divides $aF + bG$, that is, $aF(\mathbf{x}) + bG(\mathbf{x}) = H(\mathbf{x})K(\mathbf{x})$ for some K of degree $n - m$. Since all n^2 common points of $F = 0$ and $G = 0$ are on $aF + bG = 0$, the $n^2 - mn$ points not on $H = 0$ must be on $K = 0$. ▌

A special case of this theorem is a beautiful result discovered by B. Pascal (1623–1662) (see Exercise 2).

COROLLARY If A, B, C, A', B', C' are distinct points on a prime conic, points $AB' \wedge A'B$, $AC' \wedge A'C$, $BC' \wedge B'C$ are collinear.

PROOF. The three-line sets $\{AC', BA', CB'\}$ and $\{AB', BC', CA'\}$ are curves of order $n = 3$. They intersect in $n^2 = 9$ points, exactly six of which lie on the prime conic of degree $m = 2$. By the theorem, the remaining $n^2 - mn = 3$ points lie on a curve of order $n - m = 1$. ▌

This corollary is a completion of Pappus' theorem, since the two given lines in the Pappus configuration constitute a (nonprime) conic.

Exercises

1. Prove Theorem 1.
2. How old was Pascal when he discovered the result presented above as a corollary to Theorem 3?
3. Are the points of intersection in the corollary to Theorem 3 necessarily unique?
4. Generalize the corollary to Theorem 3 by allowing certain of the six points to be equal and interpreting "line PP" as "the tangent at P."
5. Find necessary and sufficient conditions for a conic to have: (a) exactly one singular point; (b) exactly one singular point of multiplicity two; (c) exactly two singular points.
6. The curves $x_1^3 - x_0x_2^2 = 0$ and $x_1^3 - x_1^2x_0 + x_0x_2^2 = 0$ each have one singular point of multiplicity two. For each curve find the point and the two tangents through it. The singular point of the former curve is called a *cusp;* that of the latter a *node*. To see the origin of these terms, draw the curves in $\mathbf{A}_2\mathrm{R}$. (Although this is a somewhat illegal procedure, since R is not algebraically closed, drawings in $\mathbf{A}_2\mathrm{R}$ are nevertheless comforting.)

*7. Prove that a prime curve of order n which has a singular point of multiplicity $n - 1$ can have no other singular points.

8. Prove the corollary to Theorem 3 by using Theorem 2.7 instead of Theorem 3.

5. CUBICS

A *cubic* is a plane curve of order three. This is a sufficiently high order for some of the more interesting properties of plane curves to appear. However, we can present here only a glimpse of these properties. The reader would do well to pursue the subject further by consulting, for example, [20] or [24].

Let $A\colon\mathbf{a}$ be a point on the curve $F = 0$ (not necessarily a cubic). Let $B\colon\mathbf{b}$ be another point such that AB is not a component of $F = 0$. Then $G(u, v) = F(u\mathbf{a} + v\mathbf{b})$ is a homogeneous polynomial of the same degree as F. The ratio $1:0$ is a root of G of some multiplicity $m > 0$ (Section I.8). Then A is said to be an *m-fold point of intersection of $F = 0$ and AB*. We shall require this concept in a moment.

THEOREM 1 Every nonsingular cubic has, in some affine coordinate system, an equation of the form $y^2 = \phi(x)$, where ϕ is a cubic polynomial with distinct roots.

PROOF. Let $F = 0$ be the cubic. The second derivatives $F_{ij}(\mathbf{x})$ are homogeneous of degree one. Let $H(\mathbf{x})$ be the determinant of the 3×3 matrix

$(F_{ij}(\mathbf{x}))$. H is homogeneous of degree three. Let A:$\mathbf{a}$ be a common point of the cubics $F = 0$ and $H = 0$ (Exercise 1). Let $\mathcal{C}$ be the conic $\sum F_{ij}(\mathbf{a})x_ix_j = 0$. Since A is on $H = 0$, the matrix $(F_{ij}(\mathbf{a}))$ is singular; therefore, $\mathcal{C}$ is singular. Applying Euler's theorem (Section 1.8), we have

$$\sum_i \Big(\sum_j F_{ij}(\mathbf{a})a_j\Big) a_i = \sum_i 2F_i(\mathbf{a})a_i = 6F(\mathbf{a}) = 0;$$

hence, A is on $\mathcal{C}$. The tangent line t_A to $\mathcal{C}$ at A has equation $\sum_{i,j} F_{ij}(\mathbf{a})a_ix_j = 0$. Again by Euler's theorem this is the same as the line with the equation, $\sum_j F_j(\mathbf{a})x_j = 0$. Since $\mathcal{C}$ is singular, this line is a component of $\mathcal{C}$ (Theorem 2.1). Thus, $\sum F_{ij}(\mathbf{a})b_ib_j = 0$ for all B:(b_0, b_1, b_2) on t_A. It follows that t_A has a threefold intersection with $F = 0$ at A.

Now choose coordinates in which A:$(0, 0, 1)$ and t_A:$[1, 0, 0]$. Since $F(A) = F_1(A) = F_2(A) = 0$ and $F_0(A) \neq 0$, $F(\mathbf{x})$ is of the form $a_{00}x_0^3 + a_{01}x_0^2x_1 + a_{02}x_0^2x_2 + a_{10}x_0x_1^2 + x_0x_2^2 + a_{11}x_1^3 + a_{12}x_1^2x_2 + ax_0x_1x_2$. Using A and the point with coordinate $(0, 1, 0)$ as a basis, all points on t_A have coordinates of the form $(0, v, u)$. $F(0, v, u) = a_{11}b^3 + a_{12}uv^2$ must have u:$v = 1$:0 as a root of multiplicity three. Hence, $a_{11} \neq 0$ and $a_{12} = 0$. In affine coordinates, then, the equation of the cubic is of the form $x^3 + ay^2 + bxy + cx^2 + dx + ey + k = 0$, with $a \neq 0$. This can be written as follows: $(2ay + bx + e)^2 = (bx + e)^2 - 4a(x^3 + cx^2 + dx + k)$. Perform the coordinate change $x \leftrightarrow x$, $y \leftrightarrow (y - bx - e)/2a$ and let $\phi(x) = (bx + e)^2 - 4a(x^3 + cx^2 + dx + k)$. Then the curve has the equation $y^2 = \phi(x)$. If $\phi(x)$ had a repeated root r, the point $(r, 0)$ would be a singular point. ∎

The special point A used above is an example of the following definition.

DEFINITION 1 A nonsingular point P of a curve is a *flex* if the tangent at P has at least a threefold intersection with the curve at P. We have already proved most of a second theorem.

THEOREM 2 A nonsingular point P of a curve $F = 0$ is a flex if and only if the matrix $(F_{ij}(P))$ is singular.

PROOF. Exercise 2.

The determinant $H(\mathbf{x})$ of $(F_{ij}(\mathbf{x}))$ is called the *Hessian* of $F(\mathbf{x})$.

We can now obtain rather full information on the flexes of cubics.

THEOREM 3 Every nonsingular cubic has exactly nine flexes. Each line through two of these flexes contains a third.

PROOF. By Theorem 1 the cubic has an affine equation of the form $y^2 = k(x - b_1)(x - b_2)(x - b_3)$, with one flex on l_∞. By the coordinate change $y \leftrightarrow k^{-1}y$, $x \leftrightarrow k^{-1}x + b_1$ this becomes

$$y^2 = x(x^2 + ax + b). \qquad (*)$$

Since the polynomial in x of (*) has distinct roots, we must have

$$b \neq 0 \quad \text{and} \quad a^2 - 4b \neq 0. \qquad (\dagger)$$

Returning to homogeneous coordinates, calculating the Hessian of (*), and coming back to affine coordinates, we obtain $(3x + a)y^2 + bx(3x + a) - (ax + b)^2$ (we have omitted a factor of -8). Substituting from (*) and applying Theorem 2, the affine flexes of the curve are precisely those points whose x-coordinate satisfies

$$3x^4 + 4ax^3 + 6bx^2 - b^2 = 0. \qquad (**)$$

The discriminant of (**) (Section 1.7) is $-12^4b^4(a^2 - 4b)^2$, which, in view of (†) is not zero. Hence, (**) has four distinct roots. None of the roots can give $y = 0$ in (*), since $x = 0 \rightarrow b = 0$ and $x^2 + ax + b = 0 \rightarrow a^2 - 4b = 0$. Hence, each root corresponds to two distinct points, and with the flex on l_∞ there are exactly nine flexes.

Given any two flexes on the cubic, we may choose coordinates in which they are $(0, 0, 1)$ and $(1, p, q)$, $q \neq 0$, and the curve has affine equation (*). Then $x = p$, $y = -q$ satisfy (*) and (**); so $(1, p, -q)$ is a third flex on the line $[p, -1, 0]$ determined by the two flexes. ∎

For results on flexes of singular cubics see the exercises.

Exercises

1. Prove that any two curves in $\mathbf{P}_2K$ have a common point.
2. Prove Theorem 2.
3. Can a conic have flexes?
4. Show that a prime cubic that has a point of multiplicity two with distinct tangents has just three flexes, and they are collinear.
5. Show that a prime cubic that has a point of multiplicity two with identical tangents has only one flex.

6. GENERALIZATIONS

Some of the results so far obtained in this chapter, for example, the theorems on conics, hold for algebraically closed fields of nonzero characteristic too. But some results, such as Theorem 4.2, do not hold for such fields without some changes in hypotheses. One reason for this difficulty is that in a field of finite characteristic p the formal derivative of the polynomial x^p is zero. Generally speaking, analytic geometry over fields of nonzero characteristic requires rather more algebraic machinery (see, for example, [2]).

In analytic geometry over fields that are not algebraically closed, even more unpleasant situations can arise. We give two examples in $\mathbf{P}_2\mathbf{R}$:

(1) The equations $x_0^2 + x_1^2 = 0$ and $x_0^4 + x_1^4 = 0$ describe the same point curve.

(2) The curve $x_1^4 - x_0^2x_1^2 - x_0^2x_2^2 = 0$ has a singular point at $(1, 0, 0)$, but it has no tangents whatsoever at this point.

Of course, some of our previous results hold over any field (and even more generally). For example, every quadric has a self-polar simplex with respect to which it has an equation of the form $\sum_{i=0}^{r} b_i x_i^2 = 0$, with all $b_i \neq 0$.

We shall restrict our discussion of general analytic geometry, usually called *algebraic geometry*, to affine space. As we have seen, the main advantage in considering a problem of analytic geometry in projective space is that by a good choice of the hyperplane at infinity the problem can be simplified in the resulting affine space. In most of the proofs of this chapter the actual calculations were carried out in affine space. Thus, there is little harm in this restriction.

* * *

Let F be an arbitrary field and let $\mathcal{S}$ be any set of points in $\mathbf{A}_nF$. If $f(\mathbf{x})$ is a polynomial in n variables over F, we write $f(\mathcal{S}) = 0$ to signify that $f(P) = 0$ for all P in $\mathcal{S}$. Consider

$$\mathfrak{S} = \{f \in F[\mathbf{x}] \mid f(\mathcal{S}) = 0\}.$$

It is easy to see that $\mathfrak{S}$ is an ideal in $F[\mathbf{x}]$ (Section I.9). We call $\mathfrak{S}$ the *ideal determined by* $\mathcal{S}$. Conversely, given an ideal $\mathfrak{A}$ in $F[\mathbf{x}]$ and a point P in $\mathbf{A}_nF$, we write $\mathfrak{A}(P) = 0$ to signify that $f(P) = 0$ for all f in $\mathfrak{A}$. The set

$$\mathcal{A} = \{\text{points } P \text{ in } \mathbf{A}_nF \mid \mathfrak{A}(P) = 0\}$$

is called the *algebraic set determined by* $\mathfrak{A}$. We have established a relation between point sets in $\mathbf{A}_nF$ and ideals in $F[\mathbf{x}]$. Before discussing how this relation is connected with our earlier ideas, we list some of its properties.

THEOREM 1

A. Suppose the ideals $\mathfrak{A}$ and $\mathfrak{B}$ in $F[\mathbf{x}]$ determine the algebraic sets $\mathcal{A}$ and $\mathcal{B}$ in $\mathbf{A}_nF$. Then: (1) $\mathfrak{A} \subset \mathfrak{B} \rightarrow \mathcal{A} \supset \mathcal{B}$; (2) $\mathfrak{A} + \mathfrak{B}$ determines $\mathcal{A} \cap \mathcal{B}$; (3) $\mathfrak{A} \cap \mathfrak{B}$ determines $\mathcal{A} \cup \mathcal{B}$; (4) $\mathfrak{A}\mathfrak{B}$ determines $\mathcal{A} \cup \mathcal{B}$; (5) the ideal determined by $\mathcal{A}$ determines $\mathcal{A}$ and contains every ideal that determines $\mathcal{A}$.

B. Suppose the algebraic sets $\mathcal{C}$ and $\mathcal{D}$ in $\mathbf{A}_nF$ determine the ideals $\mathfrak{C}$ and $\mathfrak{D}$ in $F[\mathbf{x}]$. Then $\mathcal{C} \subset \mathcal{D} \rightarrow \mathfrak{C} \supset \mathfrak{D}$.

PROOF. A. (1), (2), and (5): Exercise 1. B: Exercise 2. A. (3) and (4): Say $\mathfrak{A} \cap \mathfrak{B}$ determines $\mathcal{L}$ and $\mathfrak{A}\mathfrak{B}$ determines $\mathcal{M}$. By (1), $\mathcal{A} \subset \mathcal{L}$, $\mathcal{B} \subset \mathcal{L}$, and $\mathcal{L} \subset \mathcal{M}$. Suppose $(\mathfrak{A}\mathfrak{B})(P) = 0$ and $P \not\subset \mathcal{A}$. Then $f(P) \neq 0$ for some $f \in \mathfrak{A}$. But $(f\mathfrak{B})(P) = 0$. Hence, $\mathfrak{B}(P) = 0$, that is, $P \in \mathcal{B}$. Similarly, if $(\mathfrak{A}\mathfrak{B})(P) = 0$

and $P \notin \mathcal{B}$ then $P \in \mathcal{A}$. Thus $\mathfrak{A}\mathfrak{B}$ determines a subset of $\mathcal{A} \cup \mathcal{B}$. We now have $\mathcal{A} \cup \mathcal{B} \subset \mathcal{L} \subset \mathcal{M} \subset \mathcal{A} \cup \mathcal{B}$. ▮

Let $f(x, y) = 0$ be a point curve in $\mathbf{A}_2F$. Let $f_1^{e_1}f_2^{e_2} \cdots f_k^{e_k}$ be the factorization of f, where the f_i are distinct primes. The principal ideal (Section I.9) $\mathfrak{F}^* = f_1f_2 \cdots f_kF[x, y]$ determines the same algebraic set as the principal ideal $\mathfrak{F} = fF[x, y]$, namely, the union of the prime components of f, considered as point sets. Clearly, $\mathfrak{F} \subset \mathfrak{F}^*$. Thus, if we ignore multiple components we can identify the point curves previously studied with the algebraic sets determined by principal ideals. Of special importance in algebraic geometry is the following generalization of "prime component."

DEFINITION 1 A *variety* is an algebraic set that is not the union of a finite number of proper algebraic subsets.

For example, the algebraic set determined by $xy = 0$ is not a variety but is the union of the varieties $x = 0$ and $y = 0$. We show that this is the sort of thing that always happens.

THEOREM 2 Every algebraic set $\mathcal{A}$ in $\mathbf{A}_nF$ can be expressed in the form $\mathcal{A} = \mathcal{V}_1 \cup \mathcal{V}_2 \cup \cdots \cup \mathcal{V}_n$, where the $\mathcal{V}_i$ are varieties and $\mathcal{V}_i \not\subset \mathcal{V}_j$ whenever $i \neq j$. The expression is unique except for the order of the $\mathcal{V}_i$.

PROOF

Existence: Suppose there are algebraic sets in $\mathbf{A}_nF$ that are not the union of a finite number of varieties. Let $\sum$ be the set of all such sets. We wish to apply the minimum principal (Section I.2) to $\sum$ with the partial order $\subset$. Let $\mathcal{A}_1 \supset \mathcal{A}_2 \supset \cdots$ be a chain in $\sum$. Let $\mathfrak{A}_i$ be the ideal determined by $\mathcal{A}_i$. Then, by Theorem 1.B, $\mathfrak{A}_1 \subset \mathfrak{A}_2 \subset \cdots$. By the Hilbert basis theorem (Section I.9) there is an N such that $\mathfrak{A}_N = \mathfrak{A}_{N+1} = \cdots$. Then, by Theorem 1(5), $\mathcal{A}_N = \mathcal{A}_{N+1} = \cdots$. Thus, $\mathcal{A}_N$ is a lower bound in $\sum$ of the chain. By the minimum principle, then, $\sum$ contains a minimal element $\mathcal{L}$. Since $\mathcal{L} \in \sum$, it is not a variety; hence, by Definition 1, $\mathcal{L} = \mathcal{L}_1 \cup \mathcal{L}_2 \cup \cdots \cup \mathcal{L}_k$ where each $\mathcal{L}_i \subsetneq \mathcal{L}$ is an algebraic set. Since $\mathcal{L}$ is minimal, $\mathcal{L}_i \notin \sum$ for each i. Then, by definition of $\sum$, each $\mathcal{L}_i$ is the union of a finite number of varieties. But then $\mathcal{L}$ is the union of a finite number of varieties, contradicting $\mathcal{L} \in \sum$. Thus, every algebraic set $\mathcal{A}$ can be represented in the form $\mathcal{A} = \mathcal{V}_1 \cup \mathcal{V}_2 \cup \cdots \cup \mathcal{V}_n$, where the $\mathcal{V}_i$ are varieties. Now pick a representation of $\mathcal{A}$ in which n is as small as possible. If $\mathcal{V}_i \subset \mathcal{V}_j$ for some $i \neq j$, then $\mathcal{V}_i$ could be omitted from the representation and n would not have been the smallest possible. Hence, $\mathcal{V}_i \not\subset \mathcal{V}_j$ whenever $i \neq j$.

Uniqueness: Suppose $A = \mathcal{V}_1 \cup \mathcal{V}_2 \cup \cdots \cup \mathcal{V}_n = \mathcal{W}_1 \cup \mathcal{W}_2 \cup \cdots \cup \mathcal{W}_m$, where $\{\mathcal{V}_i\}$, $\{\mathcal{W}_j\}$ are each sets of noncomparable varieties. Then $\mathcal{V}_1 = \mathcal{V}_1 \cap \mathcal{A} = (\mathcal{V}_1 \cap \mathcal{W}_1) \cup (\mathcal{V}_1 \cap \mathcal{W}_2) \cup \cdots \cup (\mathcal{V}_1 \cap \mathcal{W}_m)$. Since $\mathcal{V}_1$ is a variety, one of the algebraic sets, say $\mathcal{V}_1 \cap \mathcal{W}_j$, is not a proper subset of $\mathcal{V}_1$, that is, $\mathcal{V}_1 = \mathcal{V}_1 \cap \mathcal{W}_j$, that is, $\mathcal{V}_1 \subset \mathcal{W}_j$. Similarly, $\mathcal{W}_j \subset \mathcal{V}_i$ for some i. But then $\mathcal{V}_1 \subset \mathcal{V}_i$ and so $i = 1$. Hence, $\mathcal{V}_1 = \mathcal{W}_j$. Similarly, every $\mathcal{V}_i$ is some $\mathcal{W}_k$ and every $\mathcal{W}_r$ is some $\mathcal{V}_s$. ∎

The position of varieties in the relation between algebraic sets and ideals established above is shown in Theorem 3.

THEOREM 3 Varieties determine prime ideals (Section I.9).

PROOF. Suppose a variety $\mathcal{V}$ in $\mathbf{A}_nF$ determines an ideal $\mathfrak{B}$ in $F[\mathbf{x}]$, which is not prime. Then there exist f and g in $F[\mathbf{x}]$ such that $f \not\in \mathfrak{B}$, $g \not\in \mathfrak{B}$ but $fg \in \mathfrak{B}$. $\mathfrak{B}$ is properly contained in the ideal $\mathfrak{B} + fF[\mathbf{x}]$, hence, this ideal determines an algebraic set $\mathcal{A}$ that is properly contained in $\mathcal{V}$. Similarly, $\mathfrak{B} + gF[\mathbf{x}]$ determines $\mathcal{B} \subsetneq \mathcal{V}$. Then $\mathcal{A} \cup \mathcal{B} \subset \mathcal{V}$. The ideal $(\mathfrak{B} + fF[\mathbf{x}])(\mathfrak{B} + gF[\mathbf{x}]) = \mathfrak{B} + f\mathfrak{B} + g\mathfrak{B} + fgF[\mathbf{x}]$ is contained in $\mathfrak{B}$, since $fg \in \mathfrak{B}$. But by Theorem 1(4) this ideal determines $\mathcal{A} \cup \mathcal{B}$. Thus, $\mathcal{V} \subset \mathcal{A} \cup \mathcal{B}$. We now have $\mathcal{V} = \mathcal{A} \cup \mathcal{B}$ where $\mathcal{A}$, $\mathcal{B}$ are proper algebraic subsets of $\mathcal{V}$; this contradicts Definition 1. ∎

Under more complicated hypotheses it can be shown that every prime ideal $\mathfrak{P}$ determines a variety, and that this variety determines $\mathfrak{P}$. Thus our relation between algebraic sets and ideals induces a one–one order reversing correspondence between prime ideals and varieties. Such is the beginning of algebraic geometry—one of the most active areas of research in mathematics today. Students wishing to pursue the subject might begin by reading [12] and [2]. The most important works are perhaps [25] and [9].

Exercises

1. Prove Theorem 1, parts A(1), (2), (5).
2. Prove Theorem 1.B.
3. Can part B of Theorem 1 be extended along the lines of (2)–(5) of part A?
4. A *generic point* of a variety $\mathcal{V}$ in $\mathbf{A}_nF$ is a point P such that $\{P\}$ determines an ideal in $F[\mathbf{x}]$ that determines $\mathcal{V}$. Find generic points for the following varieties in $\mathbf{A}_2\mathbf{C}$: (a) all of $\mathbf{A}_2\mathbf{C}$; (b) $\{(x, y) \mid x - y = 0\}$; (c) $\{(0, 0)\}$.
5. (For topology students.) Let $\mathcal{P}$ be the set of all points in $\mathbf{A}_nF$. Show that the algebraic sets in $\mathbf{A}_nF$ are the open sets of a topology for $\mathcal{P}$. What properties does this topology have?

VI

Projective Planes

The general theory of projective planes is a small but lively field of contemporary mathematics. In the following sections we shall present a number of results in this theory that can be proved without too much algebraic machinery, although some will be required. Sections I.10, I.11, and I.12 should now be read. We hope that the student will go on to the more advanced works on the subject, especially *Projektive Ebenen*, by G. Pickert, and *Finite Geometries*, by H. P. Dembowski. (See "Books for Further Study," following the bibliography.)

Throughout this chapter, *plane* shall mean *projective plane*.

1. ORDERS OF PLANES AND SUBPLANES

A *subplane* of a projective plane $\mathbf{P} = (\mathcal{P}, \mathcal{L}, \epsilon)$ is a projective plane $\mathbf{P}' = (\mathcal{P}', \mathcal{L}', \epsilon')$ such that $\mathcal{P}' \subset \mathcal{P}$, $\mathcal{L}' \subset \mathcal{L}$, and $\epsilon' = \mathcal{P}' \times \mathcal{L}' \cap \epsilon$. In other words, a subplane of $\mathbf{P}$ is a subset that is a plane under the restricted incidence relation. There is a rather weak analog for planes to Lagrange's theorem for groups (Section I.11), which will be useful later.

THEOREM 1 (R. Bruck.) Let $\mathbf{P}'$ be a proper subplane of order m of a finite plane $\mathbf{P}$ of order n. If every line of $\mathbf{P}$ contains a point of $\mathbf{P}'$, then $n = m^2$; otherwise, $n \geq m^2 + m$.

PROOF. Let l be a line of $\mathbf{P}'$. Since $\mathbf{P}'$ is a proper subplane, there is a point $P \epsilon l$ not in $\mathbf{P}'$. There are m^2 lines connecting P to the m^2 points of $\mathbf{P}'$ not on l. If two of these lines were equal, say $PQ_1 = PQ_2$, then $P = Q_1Q_2 \wedge l \epsilon P'$, a contradiction. Hence, P is on at least $m^2 + 1$ lines, and so $m^2 \leq n$. If $m^2 < n$, there is a line l' through P that contains no point of $\mathbf{P}'$. The lines of $\mathbf{P}'$ intersect l' in at most $m^2 + m + 1$ points. If two of these points were equal, say $l' \wedge l_1 = l' \wedge l_2$, then $l_1 \wedge l_2$ would be a point in $\mathbf{P}'$ on l', a contradiction. Hence, l' contains at least $m^2 + m + 1$ points, and so $m^2 + m \leq n$. ∎

What are the possible orders of finite planes? It might seem at first that there should be planes of every order $n > 1$, just as there are groups of every order. That this is not the case is shown by the following theorem, which excludes an infinite number of $n > 1$ from being orders of planes.

THEOREM 2 (R. Bruck and H. Ryser.) If $n \equiv 1 \pmod 4$ or $n \equiv 2 \pmod 4$ and if n is not the sum of two squares, then there is no projective plane of order n.

PROOF. Suppose there is a plane of order $n \equiv 1$ or $2 \pmod 4$ with points $P_1, P_2, \cdots, P_N$ and lines $l_1, l_2, \cdots, l_N$, where $N = n^2 + n + 1$. Let $x_1, x_2, \cdots, x_N$ be independent variables. For each $j = 1, 2, \cdots, N$, define the polynomial $L_j = \sum_{i=1}^{N} a_{ij}x_j$, where $a_{ij} = 1$ if $P_i \in l_j$, $a_{ij} = 0$ if $P_i \notin l_j$.
Then (Exercise 1)

$$L_1^2 + L_2^2 + \cdots + L_N^2 = n(x_1^2 + x_2^2 + \cdots + x_N^2) + (x_1 + x_2 + \cdots + x_N)^2. \quad (1)$$

Then (Exercise 2)

$$L_1^2 + L_2^2 + \cdots + L_N^2 = n\left(x_2 + \frac{x_1}{n}\right)^2 + n\left(x_3 + \frac{x_1}{n}\right)^2 + \cdots + n\left(x_N + \frac{x_1}{n}\right)^2 + (x_2 + x_3 + \cdots + x_N)^2. \quad (2)$$

Let

$$y_1 = x_2 + x_3 + \cdots + x_N, \qquad y_i = x_i + \frac{x_1}{n}, \qquad 2 \leq i \leq N. \quad (3)$$

Then (2) becomes

$$L_1^2 + L_2^2 + \cdots + L_N^2 = y_1^2 + ny_2^2 + ny_3^2 + \cdots + ny_N^2. \quad (4)$$

Since $n \equiv 1$ or $2 \pmod 4$, $N \equiv 3 \pmod 4$. Thus, if we break up the terms of (4) into groups of four starting with ny_2^2, there will be two terms left over:

$$y_1^2 + n(y_2^2 + y_3^2 + y_4^2 + y_5^2) + \cdots + n(y_{N-5}^2 + y_{N-4}^2 + y_{N-3}^2 + y_{N-2}^2) + n(y_{N-1}^2 + y_N^2).$$

From Section I.10 we know there are integers a_1, a_2, a_3, a_4 such that

$$n = a_1^2 + a_2^2 + a_3^2 + a_4^2.$$

Applying Lagrange's identity (Section I.10), each

$$n(y_i^2 + y_{i+1}^2 + y_{i+2}^2 + y_{i+3}^2) = z_i^2 + z_{i+1}^2 + z_{i+2}^2 + z_{i+3}^2,$$

where the z's are linear combinations of the y's with integer coefficients. Letting

$$z_1 = y_1, \qquad z_{N-1} = y_{N-1}, \qquad z_N = y_N,$$

(4) becomes

$$L_1^2 + L_2^2 + \cdots + L_N^2 = z_1^2 + z_2^2 + \cdots + z_{N-2}^2 + n(z_{N-1}^2 + z_N^2). \quad (5)$$

We are going to simplify (5) by judiciously introducing relations among the hitherto independent z's. Using (3) and the definition of the z's, we may write the x's as linear combinations of z's with rational coefficients.

Then we have $L_1 = r_1z_1 + r_2z_2 + \cdots + r_Nz_N$, where the r_i are rational numbers. If $r_1 \neq 1$ set $z_1 = (1 - r_1)^{-1}(r_2z_2 + \cdots + r_Nz_N)$; if $r_1 = 1$ set $z_1 = -\frac{1}{2}(r_2z_2 + \cdots + r_Nz_N)$. Then, in any case, $L_1 = \pm z_1$, and so (5) reduces to $L_2^2 + L_3^2 + \cdots + L_N^2 = z_2^2 + z_3^2 + \cdots + z_{N-2}^2 + n(z_{N-1}^2 + z_N^2)$. The remaining z's are still independent variables, so we may continue this process, forcing $L_2 = \pm z_2$, then $L_3 = \pm z_3, \cdots$, then $L_{N-2} = \pm z_{N-2}$. Equation (5) has now been reduced to

$$L_{N-1}^2 + L_N^2 = n(z_{N-1}^2 + z_N^2), \tag{6}$$

where z_{N-1}, z_N are still independent variables. L_{N-1} and L_N are linear combinations of z_{N-1} and z_N with rational coefficients, say,

$$L_{N-1} = \frac{m_1}{n_1} z_{N-1} + \frac{m_2}{n_2} z_N, \qquad L_N = \frac{s_1}{t_1} z_{N-1} + \frac{s_2}{t_2} z_N,$$

where the m_i, n_i, s_i, t_i are integers. In (6) set $z_{N-1} = n_1t_1$ and $z_N = n_2t_2$ to obtain

$$(m_1t_1 + m_2t_2)^2 + (s_1n_1 + s_2n_2)^2 = n((n_1t_1)^2 + (n_2t_2)^2).$$

Then, by Corollary 1 of Section I.10, n is the sum of two squares. ▌

Every finite *Desarguesian* plane can be coordinatized by a finite field and, hence (Section I.6), is of prime power order. Many finite non-Desarguesian planes have been found, but all known finite planes are of prime power order. The smallest $n > 1$ that are not prime powers are 6 and 10. By the above, Theorem 6 cannot be the order of a plane. However, $10 = 3^2 + 1^2$ is not excluded by this theorem. It would be a great achievement to find a plane of order 10 or to prove that no such plane exists.

Exercises

1. In the proof of Theorem 2, establish Eq. (1). (Hint: Use the first result of Exercise III.1.5.)
2. Establish (2) in the proof of Theorem 2.
3. In the proof of Theorem 2 the reduction of (5) can be accomplished only because $z_1, z_2, \cdots, z_N$ are independent. Prove they are independent. (First prove $y_1, y_2, \cdots, y_N$ are independent.)
4. Prove that the only planes of order 2 or 3 are the Desarguesian planes. (This is also true for orders 4, 5, 7, and 8, but not for 9.)

* * *

The study of curves in an arbitrary plane has not as yet yielded detailed results of the sort obtained in Chapter V. However, some results have been obtained.

5. An *arc* in a plane is a set of points no three of which are collinear. A line is *tangent* to an arc if it meets the arc in exactly one point. Prove that the total number of tangents to an arc in a finite plane is even if and only if the number of points on the arc is even.
6. (See Exercise 5.) Show that no arc in a finite plane of order n has more than $n + 1$ points, if n is odd, or $n + 2$ points, if n is even. Give examples to show that these bounds can be attained.
7. (See Exercise 5. Also, compare with Definition II.6.6.) An *oval* in a plane of finite odd (even) order n is an arc of $n + 1$ $(n + 2)$ points. Prove that no three tangents to an oval in a finite plane of odd order are concurrent. (It is not known if every finite plane contains an oval!)
8. (See Exercises 5 and 7.) Let **O** be an oval in a finite plane of odd order n. A point P not on **O** is *exterior* if it lies on a tangent; *interior* if it does not. Show that **O** has exactly $\frac{1}{2}(n + 1)$ exterior points and $\frac{1}{2}(n - 1)$ interior points.

It is known that every oval in a finite Desarguesian plane of odd order is an irreducible conic.* This suggests that our definition of oval is a good one for finite planes of odd order. The corresponding result for finite planes of even order is, however, false. An oval in a plane of infinite order is sometimes defined as an arc that has a unique tangent line through each of its points.

*9. (N. Krier.) Construct in the Euclidean plane a bounded oval that is not the boundary of a convex set.

2. POLARITIES AND INVOLUTIONS

Many results on collineations and correlations of planes make use of the theorems of this section, which concern collineations and correlations of the smallest possible order.

DEFINITION 1 A *polarity* of a plane **P** is a correlation α of **P** such that $x\alpha^2 = x$ for all points and lines x. A point or line x is an *absolute* element of α if $x\alpha$ and x are incident.

Let α be a polarity of a finite plane **P** of order n.

THEOREM 1 Every absolute line of α contains exactly one absolute point.

PROOF. If l is absolute, so is $L = l\alpha$. If $P \in l$, $P \neq L$, were also absolute, then $L \in P\alpha$ and $P \in P\alpha$ would imply $P\alpha = PL = l = L\alpha$; hence, $P = L$. ∎

---

* B. Segre, "Ovals in a finite projective plane," *Canad. J. Math.*, **7** (1955), pp. 414–416.

THEOREM 2 Every line that is not absolute contains an even number of points that are not absolute.

PROOF. If l is not absolute, then $L = l\alpha \notin l$. If $P \in l$, then $L \in P\alpha$; hence, $P\alpha \neq l$. Let $P' = P\alpha \wedge l$. P is absolute iff $P \in P\alpha$, and this occurs iff $P = P'$. Thus, the nonabsolute points of l occur in distinct pairs. ∎

COROLLARY 1 If n is even, every line contains an odd number of absolute points.

PROOF. Exercise 1.

COROLLARY 2 If n is odd, then a line is absolute if and only if it contains exactly one absolute point.

PROOF. Exercise 2.

THEOREM 3 (R. Baer.) If M is the total number of absolute points in P, then: (a) $M \equiv n + 1 \pmod 2$; (b) if p is an odd prime and j a nonnegative integer, then $(M - n - 1)n^{(p^j-1)/2}(n^{(p-1)p^j/2} - 1) \equiv 0 \pmod{p^{j+1}}$.

PROOF. An *m-cycle* (m a positive integer) is an m-tuple of points $(P(1), P(2), \cdots, P(m))$ such that

$$P(i) \in P(i + 1)\alpha, \qquad 1 \leq i \leq m - 1, \qquad \text{and } P(m) \in P(1)\alpha.$$

Let $N(m)$ be the total number of m-cycles in P. We prove three assertions about $N(m)$.

I. If $m > 2$, then $N(m) = (n^2 + n + 1)(n + 1)^{m-2} + nN(m - 2)$. By an *m-chain* we mean an $(m - 1)$-tuple of points $(P(1), \cdots, P(m - 1))$ such that $P(i) \in P(i + 1)\alpha$, $1 \leq i \leq m - 2$. We may construct an m-cycle by choosing any point for $P(1)$, any point on $P(1)\alpha$ for $P(2)$, any point on $P(2)\alpha$ for $P(3)$, and so on. Thus, there are exactly

$$(n^2 + n + 1)(n + 1)^{m-2} \tag{1}$$

m-chains. An m-chain is *short* if $P(m - 1) = P(1)$; otherwise, it is *long*. If $(P(1), \cdots, P(m - 1))$ is a short m-chain then $(P(1), \cdots, P(m - 2))$ is an $(m - 2)$-cycle. Conversely, if $(P(1), \cdots, P(m - 2))$ is an $(m - 2)$-cycle, then $(P(1), \cdots, P(m - 2), P(1))$ is a short m-chain. Thus, there are exactly

$$N(m - 2) \tag{2}$$

short m-chains. If $(P(1), \cdots, P(m - 2), P(1))$ is a short m-chain, then $(P(1), \cdots, P(m - 2), P(1), P)$ is an m-cycle iff $P \in P(1)\alpha$. Thus, there are exactly

$$(n + 1)N(m - 2) \tag{3}$$

m-cycles beginning with short m-chains. If $(P(1), \cdots, P(m - 1))$ is a long m-chain, then $P = P(1)\alpha \wedge P(m - 1)\alpha$ is the only point such that

$(P(1), \cdots, P(m-1), P)$ is an m-cycle. Thus, the number of m-cycles beginning with long m-chains = the number of long m-chains = the number of m-chains minus the number of short m-chains. Subtracting (2) from (1) and adding (3) gives I.

II. For $m \geq 0$, $N(2m+1) = (n+1)^{2m+1} + n^m(M-n-1)$. We show this by induction on m. For $m = 0$, II becomes $N(1) = M$. This is true because (P) is a one-cycle iff P is absolute. Suppose II for $m = k-1$. Then by I,

$$\begin{aligned} N(2k+1) &= (n^2+n+1)(n+1)^{2k-1} + nN(2k-1) \\ &= (n^2+n+1)(n+1)^{2k-1} + n((n+1)^{2k-1} + n^{k-1}(M-n-1)) \\ &= (n+1)^{2k+1} + n^k(M-n-1). \end{aligned}$$

III. For every prime p and integer $j > 0$, $N(p^j) \equiv N(p^{j-1}) \pmod{p^j}$. A p^{j-1}-cycle written p times in succession:

$$(\overline{P(1), \cdots, P(p^{j-1})}, \overline{\qquad\qquad\qquad}, \cdots, \overline{\qquad\qquad\qquad}) \qquad (*)$$

is a p^j-cycle. There are $N(p^j) - N(p^{j-1})$ p^j-cycles that are not of this form; Let $(P(1), \cdots, P(p^j))$ be one of these.

ASSERTION. No two of the p^j-cycles $(P(i), P(i+1), \cdots, P(p^j), P(1), \cdots, P(i-1))$, $1 \leq i \leq p^j$, are equal. For, suppose not, that is, suppose there is an i, $0 < i < p^j$, such that $P(k) = P(k+i)$ for all k (where large $k+i$ are reduced mod p^j). Let p^h be the highest p-power dividing i, so that $i = i'p^h$, $p \nmid i'$, $0 \leq h < j$. The cyclic permutation $\pi: k \rightarrow k + i \pmod{p^j}$ fixes $P(1), \cdots, P(p^j))$, hence, every power of π fixes it. By Corollary I.10.3 there is a t such that $ti' \equiv 1 \pmod{p^{j-h}}$. Then $ti \equiv p^h \pmod{p^j}$, and so π^t: $k \rightarrow k + p^h$. Thus, since π^t fixes our cycle, our cycle consists of the block $\overline{P(1), \cdots, P(p^h)}$ repeated p^{j-h-1} times. But then our cycle can be written in the form (*), a contradiction.

By the assertion, then, the $N(p^j) - N(p^{j-1})$ p^j-cycles not of the form (*) can be split into disjoint classes each containing exactly p^j p^j-cycles. Hence, $N(p^j) - N(p^{j-1}) \equiv 0 \pmod{p^j}$, and III is proved.

PROOF OF (A). By III, $M = N(1) \equiv N(2) = (n+1)(n^2+n+1) \equiv n+1 \pmod{2}$.

PROOF OF (B). If p is an odd prime, then by II and III

$$\begin{aligned} (n+1)^{p^j} + n^{(p^j-1)/2}(M-n-1) &\equiv N(p^j) \equiv N(p^{j-1}) \\ &\equiv (n+1)^{p^{j-1}} + n^{(p^{j-1}-1)/2}(M-n-1) \pmod{p^j}. \end{aligned}$$

By Corollary I.10.2,

$$(n+1)^{p^j} \equiv (n+1)^{p^{j-1}} \pmod{p^j}.$$

Hence,

$$(M - n - 1)(n^{(p^j-1)/2} - n^{(p^{j-1}-1)/2}) \equiv 0 \pmod{p^j}.$$

Replace j by $j + 1$ and rearrange to get (b). ∎

COROLLARY 3 Every polarity of a finite plane has an absolute point.

PROOF. Exercise 3.

An *involution* is a collineation of order two (Definition II.5.4). R. Baer has also proved a very useful result on involutions.

THEOREM 4 If α is an involution of a finite plane of order n, then either: (1) $n = m^2$ and the fixed points and lines of α form a subplane of order m; or (2) α is a perspectivity. In case (2): (i) if n is odd, α is a homology; (ii) if n is even, α is an elation.

PROOF. We first prove an assertion.

ASSERTION. Every point is on a fixed line (and dually). Let P be a point. If $P \neq P\alpha$, then $PP\alpha$ is fixed. If $P = P\alpha$, choose another point Q. $(PQ)\alpha = PQ\alpha$, so if PQ is not fixed, then $Q\alpha \notin PQ$. In that case, choose R on PQ, $P \neq R \neq Q$. Then $(RQ\alpha)\alpha = R\alpha Q \neq RQ\alpha$, so $S = RQ\alpha \wedge R\alpha Q$ is well-defined. $S\alpha$ is on $R\alpha Q$ and on $(R\alpha Q)\alpha = RQ\alpha$. Hence, $S\alpha = S$, and PS is fixed.

Now, if A and B are distinct fixed points, AB is fixed; dually for lines. Thus the set of fixed elements of α forms a subplane provided there are four fixed points, no three collinear.

CASE 1. There are four fixed points, no three collinear. Since α is not the identity function, the subplane of fixed elements is a proper subplane, say, of order m. By the assertion, every line of the plane meets the subplane. Hence, by Theorem 1.1, $m^2 = n$.

CASE 2. There are not four fixed points, no three collinear. We first find a line l containing three fixed points. Let l_1 be a line and $P_1 \in l_1$ a fixed point. Let l_2 be a line not through P_1 and $P_2 \in l_2$ a fixed point. Let l_3 be a line not through P_1 or P_2 and $P_3 \in l_3$ a fixed point. If $P_3 \in P_1P_2$, take $l = P_1P_2$. Otherwise, let l_4 be a line not through P_1, P_2 or P_3 and $P_4 \in l_4$ a fixed point. One of the lines P_1P_2, P_1P_3, or P_2P_3 must contain P_4; let l be that line. At most one point not on l is fixed. If $P \in l$, then, there is a line $m \neq l$ through P that contains no fixed point off l. Then P must be the fixed point of m. Hence, every point of l is fixed, and so α is a perspectivity with axis l. Since α is an involution it interchanges the nonfixed points in pairs. Therefore, the number of nonfixed points is even. (i) n even. The number of fixed points not on l is n^2 − (even number) = (even number); hence, it is zero; hence α

is an elation. (ii) n odd. The number of fixed points off l is odd; hence, it is one; hence α is a homology. ∎

Exercises

1. Prove Corollary 1.
2. Prove Corollary 2.
3. Prove Corollary 3.
4. With notation as in the proof of Theorem 3, prove for all $m \geq 1$ that $N(2m) = (n + 1)^{2m} + n^{m+1}(n + 1)$.
5. (T. G. Ostrom.) Let σ_1, σ_2 be involutary homologies with centers C_1, C_2 and axes a_1, a_2, respectively, where $C_1 \in a_2$ and $C_2 \in a_1$. Prove that $\sigma_1\sigma_2$ is an involutary homology with center $a_1 \wedge a_2$ and axis C_1C_2.

3. GROUPS OF PERSPECTIVITIES

The set P of all perspectivities of a plane **P** does not generally form a group under function composition. But, as we shall see, certain subsets of P do form groups. If **P** is of finite order, these groups can be used to obtain striking improvements of our theorem that **P** is Desarguesian if P is "as large as possible."

We begin by collecting a number of simple results on groups of perspectivities. If l is a line of **P**, denote by $P(l)$ the set of all perspectivities with axis l and by $E(l)$ the subset of all elations with axis l. If C is a point on l, let $E(C, l)$ be the set of all (C, l)-elations. If D is a point not on l, let $H(D, l)$ be the set of all (D, l)-homologies.

THEOREM 1

(a) $P(l)$, $E(l)$, $E(C, l)$, and $H(D, l)$ are groups under function composition.

(b) If μ is any collineation, $\mu^{-1}H(D, l)\mu = H(D\mu, l\mu)$, $\mu^{-1}E(C, l)\mu = E(C\mu, l\mu)$, and $\mu^{-1}E(l)\mu = E(l\mu)$.

(c) If $H(D, l)$ contains more than the identity, and G is any subgroup of $P(l)$ containing $H(D, l)$, then $N_G(H(D, l)) = H(D, l)$.

PROOF

(a) Exercise 1.

(b) If $\sigma \in H(D, l)$ and m is a line through D, $(m\mu)\mu^{-1}\sigma\mu = (m\sigma)\mu = m\mu$. Hence, $D\mu$ is the center for $\mu^{-1}\sigma\mu$. Similarly, $l\mu$ is the axis. Thus, $\mu^{-1}H(D, l)\mu \subset$

$H(D\mu, l\mu)$, and so they are equal. Similar proofs are true for $E(C, l)$ and $E(l)$.
(c) Exercise 2. ▎

The next result will be required in a later proof.

THEOREM 2 (J. André.) Let l be a line of a finite plane and C_1, C_2 points not on l. Suppose that there exist nonidentity (C_i, l)-homologies σ_i for $i = 1, 2$. Then the group G generated by $\{\sigma_1, \sigma_2\}$ contains an elation (obviously with axis l) that sends C_1 to C_2.

PROOF. Let $T = G \cap E(l)$, $g = |G|$ and $t = |T|$. Think of G and T as permutation groups on the set $\mathcal{A}$ of all points in the plane that are not on l (Section I.12). Let $\mathcal{A}_1, \mathcal{A}_2, \cdots, \mathcal{A}_r$ be the G-classes of $\mathcal{A}$. Each $\mathcal{A}_i$ is further split into, say, u_i T-classes. Each T-class contains exactly $t = |T|$ elements, since no two elements of $E(l)$ can have the same effect on a point of $\mathcal{A}$ (why?). Hence, $|\mathcal{A}_i| = u_i t$, and if $N = |\mathcal{A}|$, then

$$N = \sum_{i=1}^{r} u_i t. \tag{1}$$

If P and Q are in the same $\mathcal{A}_i$, then $P = Q\mu$ for some μ in G. Then $H(P, l) = H(Q\mu, l) = \mu^{-1}H(Q, l)\mu$. Therefore, every $H(P, l)$, for P in a given $\mathcal{A}_i$, has the same order, say, h_i. Now G is the union of T and all the $H(P, l)$, and no two of these groups have more than the identity in common (why?). Hence,

$$g = t + \sum_{i=1}^{r} u_i t(h_i - 1). \tag{2}$$

ASSERTION. For every i,

$$g = u_i h_i t. \tag{3}$$

Suppose $h_i > 1$. If $Q \in \mathcal{A}_i$, then by Theorem 1(c), $g/h_i = |G|/|N_G(H(Q, l))|$. By Section I.11 this equals the number of classes $\mu^{-1}H(Q, l)\mu$ in G. By Theorem 1(b) this equals the number of points $Q\mu$, $\mu \in G$, that is, the number $|\mathcal{A}_i| = u_i t$. Thus $g/h_i = u_i t$, hence (3). Suppose $h_i = 1$. If $Q\mu = Q\mu'$, then $\mu'\mu^{-1} \in H(Q, l)$. Then, since $h_i = 1$, $\mu'\mu^{-1} = 1$ and $\mu = \mu'$. Then $|\mathcal{A}_i| = g$ and we have $g = u_i t$, which is (3) in this case.

Now let R be total number of T-classes in $\mathcal{A}$. Then $R = N/t = \sum_{i=1}^{r} u_i$. From (1), (2), and (3),

$$g(r - 1) = t(R - 1). \tag{4}$$

If $R = 1$ then $r = 1$, that is, G is transitive on (Section I.12). Then the theorem is clear. Now suppose $R > 1$. We must show that points C_1 and C_2 in the hypothesis are in the same $\mathcal{A}_i$. Since there are nonidentity (C_i, l)-homologies, C_1 and C_2 are in classes $\mathcal{A}_i$ for which $h_i > 1$. Let r' be the number of i's for which $h_i > 1$. If we show $r' = 1$, then C_1 and C_2 must be in the same $\mathfrak{A}_i$. We may assume $h_i > 1$ for $1 \le i \le r'$ and $h_i = 1$ for $r' < i \le r$. Then, using (3) and (4),

$$\sum_{i=1}^{r'} \frac{1}{h_i} + (r - r') = \sum_{i=1}^{r} \frac{1}{h_i} = (r-1)\frac{R}{R-1} > r - 1.$$

Hence,

$$\sum_{i=1}^{r'} \frac{1}{h_i} > r' - 1.$$

But then

$$\frac{r'}{2} = \sum_{i=1}^{r'} \frac{1}{2} \ge \sum_{i=1}^{r'} \frac{1}{h_i} > r' - 1,$$

and so $r' < 2$. ∎

COROLLARY 1 Let l and m be distinct lines of a finite projective plane **P**. If for every point D on l different from $m \wedge l$ there is a nonidentity (D, m)-homology, then **P** is $(m \wedge l, m)$-transitive.

PROOF. Exercise 3.

Our first main result concerns elations.

THEOREM 3 (A. M. Gleason and A. Wagner.) If every point in a finite plane **P** is the center of a nonidentity elation and every line is the axis of a nonidentity elation, then **P** is Desarguesian.

PROOF. We shall usually omit the word "nonidentity." We first establish four statements.

I. If l is a line and P a point on l, there is a (P, l)-elation. Suppose there is no (P, l)-elation for some $P \in l$. Let $C_1, \cdots, C_s$ be those points on l that are centers for elations with axis l. Let $a_1, a_2, \cdots, a_t$ be those lines through P that are axes for elations with center P. By assumption, $P \ne C_i$ and $l \ne a_j$. Let λ_i be a (C_i, l)-elation; let μ_j be a (P, a_j)-elation. By Theorem 1(b), $\lambda_i^{-1}\mu_1\lambda_i$ is a $(P, a_1\lambda_i)$-elation; hence, $a_1\lambda_i = a_k$ for some k. We know $k \ne 1$ because l is the only line through P that is fixed by λ_i. If $a_1\lambda_i = a_1\lambda_j$, then $\lambda_i\lambda_j^{-1}$ is a (P, l)-elation, and so $\lambda_i = \lambda_j$. We have established a one–one function from $\{C_1, \cdots, C_s\}$ into $\{a_2, \cdots, a_t\}$; hence, $s < t$. But by a dual argument $t < s$, a contradiction.

II. $E(l)$ is an Abelian group in which all nonidentity elements are of the same order p, where p is a prime. Let C_1 and C_2 be distinct points of a line l. By I there is a nonidentity (C_i, l)-elation α_i for $i = 1, 2$. Let P be any point not on l. Let $l_1 = C_1P$, $l_2 = C_2P$. Then $P\alpha_1 \in l_1$; $P\alpha_2 \in l_2$; C_1, $P\alpha_2$, and $P\alpha_1\alpha_2$ are on $l_1\alpha_2$; C_2, $P\alpha_1$ and $P\alpha_2\alpha_1$ are on $l_2\alpha_1$; $P\alpha_1\alpha_2 \in C_2P\alpha_1$; $P\alpha_2\alpha_1 \in C_1P\alpha_2$. Since $C_1P\alpha_2 \ne C_2P\alpha_1$, then $P\alpha_1\alpha_2 = C_2P\alpha_1 \wedge C_1P\alpha_2 = P\alpha_2\alpha_1$. This holds for all P; therefore, $\alpha_1\alpha_2 = \alpha_2\alpha_1$. Thus, elements of $E(l)$ with different centers commute. Now suppose that α_1, β_1 have the same center. Choose α_2 with a different center. Then β_1 and $\alpha_1\alpha_2$ have different centers (why?). Hence, $\beta_1\alpha_1\alpha_2 = \alpha_1\alpha_2\beta_1 = \alpha_2\alpha_1\beta_1 = \alpha_1\beta_1\alpha_2$, and so $\beta_1\alpha_1 = \alpha_1\beta_1$. Thus $E(l)$ is Abelian. By Section I.11, $E(l)$ has an element α_1 of some prime order p. If α_2 is any element with a different center, then $(\alpha_1\alpha_2)^p = \alpha_2{}^p$ have different centers. Hence, $\alpha_2{}^p = 1$, and α_2 is of order p. We can now use α_2 to show that every element with center that of α_1 has order p. Thus, all nonidentity elements of $E(l)$ have order p.

III. $|E(C, l)|$ is the same for every C on l. Let C_1, C_2 be points on l. Choose another point P on l. Choose a line m through P that is distinct from l. By I and II, $E(P, m)$ contains an element η of prime order p. η moves every point of l except P. By Lemma I.12.1, then, $E(P, m)$ is transitive on l. Thus $E(P, m)$ contains an element ρ such that $C_1\rho = C_2$. Then $\rho^{-1}E(C_1, l)\rho = E(C_2, l)$; hence, $|E(C_1, l)| = |E(C_2, l)|$.

IV. **P** is l-transitive for every line l. By III, every $|E(P, l)| = k > 1$. Let n be the order of **P**. Since $E(l) = \bigcup_{P \in l} E(P, l)$ and distinct $E(P, l)$ have only the identity in common, $|E(l)| = t = (n + 1)(k - 1) + 1$. Since $k > 1$, $t > n$. Each $E(l)$-class of points not on l contains exactly t points. Thus, t divides the number of points not on l, that is, $n^2 = tm$ for some integer m. Since $t > n$, $n > m$. But $n^2 = [(n + 1)(k - 1) + 1]m \equiv m \pmod{n + 1}$ and $n^2 \equiv 1 \pmod{n + 1}$, and, therefore, $m = 1$. Thus, $n^2 = t$, that is, $E(l)$ is transitive on points not on l; thus, **P** is l-transitive.

Then, by Exercise III.5.4, **P** is Desarguesian. ▮

We next present a similar result on homologies.

THEOREM 4 (J. André and A. Wagner.) If every point in a finite plane **P** is the center of a nonidentity homology and every line is the axis of a nonidentity homology, and if no point or line is fixed by every collineation of **P**, then **P** is Desarguesian.

PROOF. As before, we shall usually omit the word "nontrivial." A point C is a *center for* a line l, and l is an *axis for* C, if **P** has a (C, l)-homology. We first establish six statements.

I. Either some point has more than one axis or some line has more than one center. Suppose not. Then we may define a function α as follows: If P is a point, $P\alpha$ = axis for P; if l is a line, $l\alpha$ = center for l. We show α is a polarity. Clearly, α^2 is the identity. Let $P \in l$ and let λ be a $(P, P\alpha)$-homology. Since $l\alpha$ is the center for l, $l\alpha\lambda$ is the center for $l\lambda$. But $P \in l$, so $l\lambda = l$. Therefore, $l\alpha\lambda = l\alpha$. Now, λ fixes only P and the points on $P\alpha$. $l\alpha \neq P$, since $P \in l$ and $l\alpha \notin l$; hence, $l\alpha \in P\alpha$. Then, by Corollary 3.1, there is a point A such that $A \in A\alpha$. But this is absurd, since there must be an $(A, A\alpha)$-homology.

II. If some line has more than one center, then either some line has noncollinear centers or some point has nonconcurrent axes. Suppose l has centers $C_1 \neq C_2$ but no line has noncollinear centers. Let $m = C_1C_2$, $E = l \wedge m$. Choose $D \in l$, $D \neq E$. Let λ be a homology with center D. Then $l\lambda = l$ and so $C_1\lambda$ and $C_2\lambda$ are also centers for l. Since all centers for a line are collinear, $C_1\lambda \in m$ and $C_2\lambda \in m$. Then $m\lambda = m$ and so, since $D \notin m$, m is the only axis for D. This is true for every $D \neq E$ on l. Then, by Corollary 1, **P** is (E, m)-transitive. Now there are at least two points distinct from E on l. Thus we may interchange the roles of l and m above and prove that P is also (E, l)-transitive. Then the group generated by all (E, l)- and (E, m)-elations is transitive on the set of all lines of **P** that do not contain E. Therefore, since *some* line not through E is an axis for E, *every* line not through E is an axis for E. Thus, E has nonconcurrent axes.

III. If a line l_1 has noncollinear centers, then there is a (D, l_1)-elation for every D on l_1. Suppose l_1 has noncollinear centers C_1, C_2, C_3. Let $D_1 = l_1 \wedge C_1C_3$, $D_2 = l_1 \wedge C_2C_3$. Then $D_1 \neq D_2$. By Theorem 2 there exists a (D_1, l_1)-elation ρ_1 and a (D_2, l_1)-elation ρ_2. Let D be any point on l_1 other than D_1, D_2. Let m be an axis for D. Either $D_1 \notin m$ or $D_2 \notin m$; say, $D_1 \notin m$. $m\rho_1$ is an axis for $D\rho_1 = D$ and, since D_1 is not on m, $m\rho_1 \neq m$. Then, by the dual of Theorem 2, there is an elation μ with center D such that $m\mu = m\rho_1$. Since $m \wedge m\rho_1 \in l_1$, the axis of μ is l_1.

IV. Regard the group G of all collineations of **P** as a permutation group on the set of all lines of **P**. Then every G-class contains nonconcurrent lines. Let h_1 be any line. Since G fixes no line, there is an $h_2 \neq h_1$ in the same G-class as h_1. Let $P = h_1 \wedge h_2$. There is a collineation μ such that $P\mu \neq P$. Since $h_1\mu$ and $h_2\mu$ both pass through $P\mu$, they cannot both pass through P. Thus, at least one of the triples h_1, h_2, $h_1\mu$ and h_1, h_2, $h_2\mu$ is nonconcurrent.

V. If some line has noncollinear centers, then every line is the axis of an elation. Suppose l_1 has noncollinear centers. Let G_1 be the G-class containing l_1. If every pair of lines in G_1 were concurrent with l_1, then every triple of lines in G_1 would be concurrent, contradicting IV. Hence, there are lines l_2 and l_3 in G_1 such that l_1, l_2, l_3 are nonconcurrent. Then by III and Theorem 1(b),

$$\text{there is a } (D, l_i)\text{-elation for every } D \in l_i, \qquad i = 1, 2, 3. \qquad (*)$$

Now let m be any line and C a center for m. C cannot lie on every l_i; say, $C \notin l_1$. Let $D = m \wedge l_1$ (if $m = l_1$ choose any D on l_1). By (*) there is a (D, l_1)-elation τ. Since $D \in m$, $m\tau = m$; since $C \notin l_1$, $C\tau \neq C$. Thus, there exists a (CP, m)-homology, and so, by Theorem 2, there exists a (D, m)-elation.

VI. If some line has noncollinear centers, then every point is the center of an elation. Let l_1, l_2, l_3 be as in V. Let $E_1 = l_2 \wedge l_3$, $E_2 = l_1 \wedge l_3$, $E_3 = l_1 \wedge l_2$. Let m be any line through one of the E_i; say, $E_1 \in m$. Let D be a center for m. D cannot lie on both l_2 and l_3; say, $D \notin l_2$. Let τ be an (E_1, l_2)-elation. Then

$D_T \neq D$ and $m_T = m$; hence, there is a (D_T, m)-homology. Then, by Theorem 2, there is an elation with axis m. We have found noncollinear points E_1, E_2, E_3 such that there is an (E_i, m)-elation whenever $E_i \in m$, $i = 1, 2$, or 3. This is the dual of (*). Then, by the dual of the argument following (*), we have VI.

It is now easy to prove the theorem.

CASE 1. Some line has two centers. If some line has noncollinear centers, we have V and VI. Then, by Theorem 3, **P** is Desarguesian. If no line has noncollinear centers, then, by II, some point has nonconcurrent axes. Then, using the dual of III, we have the duals of V and VI, that is, VI and V. Then, again by Theorem 3, **P** is Desarguesian.

CASE 2. No line has two centers. Then, by I, some point has two axes, and, by the dual of Case 1, **P** is Desarguesian. ▮

Most of these results are used in the following.

THEOREM 5 (A. Wagner and T. G. Ostrom.) If the collineation group of a finite plane **P** is doubly transitive (Section I.12) on the points of **P**, then **P** is Desarguesian.

The proof of this result uses some slightly more advanced group theory. It will not be given here (see [16]).

Exercises

1. Prove Theorem 1(a).
2. Using Theorem 1(b) and recalling that a nonidentity perspectivity has only one center, prove Theorem 1(c).
3. Prove Corollary 1.
4. (J. André.) Let l be a line of a finite plane **P**. Show that if more than one third of the points of **P** not on l are centers of nonidentity homologies with axis l, then **P** is l-transitive.

4. DESARGUES' CONFIGURATION AND TRANSITIVITY (CONTINUED)

We know there are non-Desarguesian planes. Any plane coordinatized by a linear planar ternary ring that is not a division ring is non-Desarguesian. The Moulton plane (Exercise III.2.4) is non-Desarguesian. Of course, these planes may contain some nontrivial Desargues' configurations; the Moulton plane certainly does. We shall now construct a *free* plane, that is, a plane in which there are no nontrivial cases of Desargues' configuration.

Let $\mathcal{P}_0$ be a set of four points. Join these points in pairs to obtain a set $\mathcal{L}_0$ of six lines. To each pair of lines of $\mathcal{L}_0$ that do not yet intersect in a point of $\mathcal{P}_0$ add a new point; let $\mathcal{P}_1$ be the set of these new points. Join each pair of points of $\mathcal{P}_0 \cup \mathcal{P}_1$ that are not yet joined by a new line; let $\mathcal{L}_1$ be the set of these new lines (Figure VI.4.1). Continuing in this way, form the sets

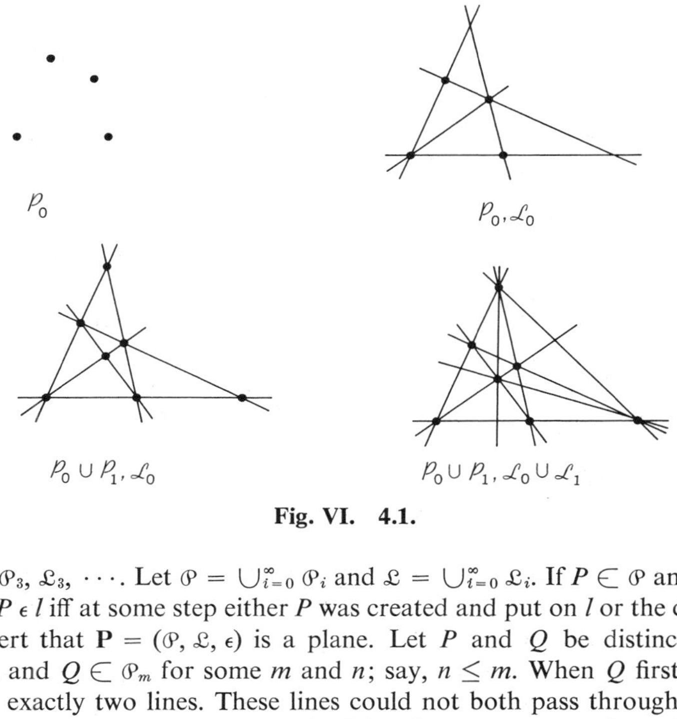

Fig. VI. 4.1.

$\mathcal{P}_2, \mathcal{L}_2, \mathcal{P}_3, \mathcal{L}_3, \cdots$. Let $\mathcal{P} = \bigcup_{i=0}^{\infty} \mathcal{P}_i$ and $\mathcal{L} = \bigcup_{i=0}^{\infty} \mathcal{L}_i$. If $P \in \mathcal{P}$ and $l \in \mathcal{L}$, we say $P \epsilon l$ iff at some step either P was created and put on l or the converse. We assert that $\mathbf{P} = (\mathcal{P}, \mathcal{L}, \epsilon)$ is a plane. Let P and Q be distinct points. $P \in \mathcal{P}_n$ and $Q \in \mathcal{P}_m$ for some m and n; say, $n \leq m$. When Q first appears it is on exactly two lines. These lines could not both pass through P, since they did not previously intersect. If neither line on Q passes through P, then P and Q will be joined by a unique line in $\mathcal{L}_m$. Thus, $\mathcal{L}_{m-1} \cup \mathcal{L}_m$ contains a unique line joining P and Q. By the construction, no further line joining P and Q will be created. Similarly, we show that any two distinct lines have a common point. Finally, no three of the four points in $\mathcal{P}_0$ are collinear, since they are on six distinct lines.

Now suppose that a nontrivial case of Desargues' configuration (Figure III.2.1) occurs in $\mathbf{P}$. Call the configuration Δ. Let n be the largest integer such that $\mathcal{P}_n$ contains a point of Δ. Let A be a point of Δ in $\mathcal{P}_n$. A is on exactly two lines of $\mathcal{L}_0 \cup \mathcal{L}_1 \cup \cdots \cup \mathcal{L}_{n-1}$; hence, at least one of the three lines of Δ that pass through A has not yet been created. Let a be such a line. Since

a has not been created, at least two of the three points of Δ that are on a are not in $\mathcal{P}_0 \cup \mathcal{P}_1 \cup \cdots \cup \mathcal{P}_{n-1}$. But all points of Δ are in $\mathcal{P}_0 \cup \mathcal{P}_1 \cup \cdots \cup \mathcal{P}_n$; hence, there is a point $B \neq A$ of Δ on a in $\mathcal{P}_n$. Let C be the third point of Δ on a. A, B, and C are not on a line in $\mathcal{L}_0 \cup \mathcal{L}_1 \cup \cdots \cup \mathcal{L}_{n-1}$, but they are in $\mathcal{P}_0 \cup \mathcal{P}_1 \cup \cdots \cup \mathcal{P}_n$. But then when $\mathcal{L}_n$ is created, A, B, and C will be joined by three distinct lines, not by the one line a. This contradiction shows that **P** is totally non-Desarguesian; in fact, it shows more (Exercise 2).

The free plane just constructed is obviously of infinite order. Although examples of non-Desarguesian planes of finite order can easily be given (assuming rather more algebraic background than we do),* they cannot be totally non-Desarguesian. For we have another theorem.

THEOREM 1 Every finite plane of order greater than three contains a nontrivial Desargues configuration.

PROOF. Let **P** be a finite plane of order greater than three. Let P be a point and l a line not through P. Let l_1, l_2, and l_3 be distinct lines through P. Since l contains at least five points, we may choose distinct points Q and R on l different from $l_1 \wedge l$, $l_2 \wedge l$, and $l_3 \wedge l$. Define a function α from $\bar{l}_1 - \{P, l_1 \wedge l\}$ into $\bar{l}$ as follows (Figure VI.4.2): $X\alpha = [(XQ \wedge l_2)R \wedge l_3] \wedge l$.

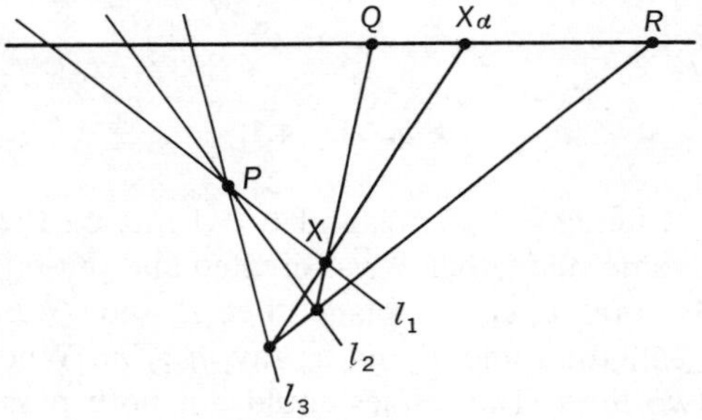

Fig. VI.4.2.

The reader should check that $X\alpha$ is well-defined for all X in $\bar{l}_1 - \{P, l_1 \wedge l\}$ and that $X\alpha$ is never equal to Q, R, $l_1 \wedge l$ or $l_3 \wedge l$. Thus, α is defined for $n - 1$ points but takes on no more than $n - 3$ values. Hence, there are distinct points A_1 and B_1 in $\bar{l}_1 - \{P, l_1 \wedge l\}$ such that $A_1\alpha = B_1\alpha$. Let $A_2 = Q_1 \wedge l_2$, $B_2 = Q_1 \wedge l_2$, $A_3 = RA_2 \wedge l_3$ and $B_3 = RB_2 \wedge l_3$. Then

---

* A most readable account of the construction of finite non-Desarguesian planes is given in Chapter 7 of *An Introduction to Finite Projective Planes* by R. Sandler and A. A. Albert.

$A_1, A_2, A_3, B_1, B_2, B_3, P, Q, R, A_1\alpha$ are the points of a nontrivial Desargues configuration. ▌ For more on this result, see Exercises 3, 4, and 5.

The set of all points or lines in a plane that are centers or axes of nontrivial Desargues' configurations can take many forms, ranging from the empty set to the entire plane. If we consider only the set of those point-line pairs (P, l) for which the plane is (P, l)-Desarguesian, we can obtain some very precise information about its possible forms. This was done by H. Lenz [13] for sets of pairs (P, l) where $P \in l$, and then in general by A. Barlotti [4].

THEOREM 2 Let $\mathfrak{F}$ be the set of all point-line pairs (P, l) of a plane $\mathbf{P}$ for which $\mathbf{P}$ is (P, l)-Desarguesian. Then $\mathfrak{F}$ must be of one of the following forms.

I.1. $\emptyset$
2. $\{(A, a)\}$, where $A \notin a$.
3. $\{(A_1, a_1), (A_2, a_2)\}$, where $A_1 \in a_2$, $A_2 \in a_1$, $A_1 \notin a_1$, and $A_2 \notin a_2$.
4. $\{(A_1, a_1), (A_2, a_2), (A_3, a_3)\}$, where $A_i \notin a_i$, $i = 1, 2, 3$, and $A_i \in a_j$ for $i \neq j$.
5. $\{(X, X\phi \mid X \in l\}$, where l is a line, L a point not on l, and ϕ a one–one correspondence between $\tilde{l}$ and $\tilde{L}$ such that, for every $X \in \tilde{l}$ and $x \in \tilde{L}$, $X \notin XQ$ and $X \in x \to \phi^{-1}(x) \in \phi(X)$.
6. $\{(X, X\psi) \mid X \in l, X \neq M\}$, where l is a line, M a point on l, and ψ a one–one correspondence between $\tilde{l} - \{M\}$ and $\tilde{M} - \{l\}$.
7. The set of I.5 together with (L, l).
8. $\{(X, X\pi) \mid X \in \mathbf{P}\}$, where π is a polarity of $\mathbf{P}$.

II.1. $\{(A, a)\}$, where $A \in a$.
2. $\{(A, a), (B, b)\}$, where $A \neq B$, $a \neq b$, $B \in a$, and $A = a \wedge b$.
3. The set of I.6 together with (M, l).

III.1. $\{(X, XL) \mid X \in l\}$, where l is a line and L a point not on l.
2. The set of III.1 together with (L, l).

IV.A.1. $\{(X, l) \mid X \in l\}$, where l is a line.
2. The set of IV.A.1 together with $\{(A, x) \mid B \in x\} \cup \{(B, x) \mid A \in x\}$, where A and B are distinct points on l.
3. $\{(X, x) \mid X \in l \text{ and } X\theta \in x\}$, where l is a line and θ a one–one correspondence between $\tilde{l}$ and itself such that θ^2 is the identity and $X\theta \neq X$ for all X on l.

IV.B. 1, 2, 3. The duals of IV.A.

V. $\{(X, l) \mid X \in l\} \cup \{(L, x) \mid L \in x\}$, where L is a point, l is a line, and $L \in l$.

VI.A. $\{(l \wedge x, x) \mid x \text{ is a line}\}$, where l is a line.

VI.B. The dual of VI.A.

VII.1. $\{(X, x) \mid X \in x\}$.
2. $\{(X, x) \mid X \text{ a point}, x \text{ a line}\}$.

PROOF. We shall prove only Lenz's parts of the theorem. The proofs of Barlotti's parts are similar but more complex. We are therefore concerned with Cases I.1, II.1, III.1, IV.A.1, IV.B.1, V, VI.A, VI.B, and VII.1. It suffices to show that if $\mathfrak{F}$ contains the set described in one of these cases together with another pair (B, b) with $B \in b$, then $\mathfrak{F}$ must contain the set described in another of these cases. For $\mathfrak{F}$ cannot contain more than the set in Case VII.1. We shall proceed by cases, omitting all dual results. Two results from previous work will be required: (α) (Theorem III.4.2) if **P** is (A, AB)- and (B, AB)-Desarguesian, then **P** is AB-Desarguesian; (β) (Exercise III.3.4) if **P** is (P, l)-Desarguesian and μ is a collineation of **P**, then **P** is $(P\mu, l\mu)$-Desarguesian.

CASE 1. $\mathfrak{F}$ contains II.1 and (B, b). If $A = B$, then $a \neq b$ and by (α) $\mathfrak{F}$ contains IV.B.1. If $a \neq b$ and $A \neq B$, then by (β) $\mathfrak{F}$ contains III.1.

CASE 2. $\mathfrak{F}$ contains III.1 and (B, b). If $L \in b$ and $B \notin l$, then by (α) every point on AB is in $\mathfrak{F}$. Then by (β) every point of every line through L is in $\mathfrak{F}$, and so $\mathfrak{F}$ contains VI.B. If $B \notin l$ and $L \notin b$, then by (β) the point $D = b \wedge l$ is on more than one line m such that (D, m) is in $\mathfrak{F}$. Hence, by (α) every line through D is in $\mathfrak{F}$, and $\mathfrak{F}$ contains V.

CASE 3. $\mathfrak{F}$ contains IV.A.1 and (B, b). If $B \in l$, then by (α) $\mathfrak{F}$ contains V. If $B \notin l$, let $C = b \wedge l$. By (β) every pair (P, m), where $P \in m$, $C \in m$ and $m \neq b$, is in $\mathfrak{F}$. Applying (β) again, every (P, b), where $P \in b$, is in $\mathfrak{F}$. Then $\mathfrak{F}$ contains IV.B.1.

CASE 4. $\mathfrak{F}$ contains V and (B, b). If $B \in l$, then by (β) $\mathfrak{F}$ contains VI.A. If $B \notin l$, then by the argument of Case 3 $\mathfrak{F}$ contains VI.B.

CASE 5. $\mathfrak{F}$ contains VI.A and (B, b). Then $b \neq l$ and $B \neq b \wedge l$. By (β) (P, b) is in $\mathfrak{F}$ for all $P \in b$ different from $b \wedge l$. Again, by (β) b can be moved to any line except l. Then $\mathfrak{F}$ contains VII.1. ∎

The complete proof of Theorem 2 requires, in addition to (α) and (β), Exercises III.4.7, 10, and 11.

Not all of the 24 forms given in Theorem 2 are actually possible. In other words, it has been shown that if $\mathfrak{F}$ contains the set described in certain of the cases then it must contain more. There are, for example, no planes of type VI.A or VI.B [10]. It has recently been shown that there are no finite planes of type I.6 [15], [26]. It is not known whether infinite planes of this type exist.

Exercises

1. Show how the process used to construct a free plane can be used to imbed any partial plane in a projective plane. Under what conditions is the plane thus obtained a free plane?

2. Show that the free plane constructed in this section also contains no nontrivial Pappus' configuration. Describe the general class of figures, containing the configurations of Desargues and Pappus, which are not in this free plane.
3. In the proof of Theorem 1, show how the result can still be obtained if $P \in l$.
4. Are there nontrivial Desargues' configurations in the planes of orders two or three?
5. Using the proofs of Theorem 1 and Exercise 3, estimate the smallest number of nontrivial Desargues' configurations that can occur in a finite plane of order n.

*6. In which Lenz–Barlotti class is the Moulton plane?

5. FANO PLANES

We recall that a plane is *Fano* if, given any four of its points P, Q, R, S, no three collinear, the diagonal points $PQ \wedge RS$, $PR \wedge QS$, and $PS \wedge QR$ are collinear. The purpose of this section is to prove the following.

THEOREM 1 (A. M. Gleason.) Every finite Fano plane is Desarguesian.

Some special group-theoretic lemmas, not given in Chapter I, will be established along with the geometrical lemmas.

Let **P** be a finite plane of order n (not necessarily a Fano plane). Let P be a point and l a line of **P**, with $P \in l$. Let $m_1, m_2, \cdots, m_n$ be the lines through P besides l. Given a point $A \neq P$ on l, denote by $\pi_{ji}{}^A$ the projection of m_i onto m_j with center A. Obviously, $\pi_{ij}{}^A\pi_{ji}{}^A$ is the identity on m_i, and $\pi_{kj}{}^A\pi_{ji}{}^A = \pi_{ki}{}^A$. Let $B \neq P$ be a point on l different from A. Then $\pi_{ij}{}^A\pi_{ji}{}^B$ is a permutation of $\tilde{m}_i$. Let $J_i^{AB} = \{\pi_{ij}{}^A\pi_{ji}{}^B \mid 1 \leq j \leq n\}$.

LEMMA 1 If Q and R are points on m_i that are different from P, then there is a unique element of J_i^{AB} that sends Q to R.

PROOF. Exercise 1.

LEMMA 2 For each η in $E(P, l)$ (Section 3) let $\bar{\eta}$ be η restricted to $\tilde{m}_i$. Then the function $\eta \longrightarrow \bar{\eta}$ is a one–one correspondence between $E(P, l)$ and the set of all permutations of $\tilde{m}_i$ that commute with every element of J_i^{AB} for all A and B, different from P, on l.

PROOF. We leave to the reader (Exercise 2) the proof that $\eta \longrightarrow \bar{\eta}$ is one–one. Suppose $\eta \in E(P, l)$, let X be any point on m_i, and let $\pi_{ji}{}^A(X) = Y$. A, X, Y are collinear; so $\eta(A), \eta(X), \eta(Y)$ are collinear. Then, since $\eta(A) = A$, $\eta(X) \in m_i$ and $\eta(Y) \in m_j$, $\pi_{ji}{}^A(\eta(X)) = \eta(Y) = \eta(\pi_{ji}{}^A(X))$. Hence, η commutes with $\pi_{ji}{}^A$, and, therefore, $\bar{\eta}$ commutes with $\pi_{ij}{}^A\pi_{ji}{}^B$. Conversely, suppose β is a permutation of $\tilde{m}_i$ that commutes with every $\pi_{ij}{}^A\pi_{ji}{}^B$. If $A \neq B$ and $i \neq j$, the only fixed point of $\pi_{ij}{}^A\pi_{ji}{}^B$ is P; hence, β must fix P. Pick a point A on l different from P and define a function α on the points of P as follows: $\alpha(X) = X$ for all X on l; if $X \in m_j$, $\alpha(X) = \pi_{ji}{}^A\beta\pi_{ij}{}^A(X)$. Clearly, α is one–one and onto α fixes all points on l and lines through P, and α restricted to $\tilde{m}_i$ is β. Thus, if we show that α preserves collinear points we will have $\alpha \in E(P, l)$ and $\bar{\alpha} = \beta$, as desired. It suffices to show that if $B \in l$, $X \in m_j$, $Y \in m_k$ are collinear, then so are $\alpha(B) = B$, $\alpha(X)$, $\alpha(Y)$. We have $Y = \pi_{kj}{}^B(X)$ and

$$\begin{aligned}\pi_{kj}{}^B\alpha(X) &= \pi_{kj}{}^B\pi_{ji}{}^A\beta\pi_{ij}{}^A(X) = \pi_{ki}{}^A(\pi_{ik}{}^A\pi_{ki}{}^B)(\pi_{ij}{}^B\pi_{ji}{}^A)\beta\pi_{ij}{}^A(X)\\ &= \pi_{ki}{}^A\beta(\pi_{ik}{}^A\pi_{ki}{}^B)(\pi_{ij}{}^B\pi_{ji}{}^A)\pi_{ij}{}^A(X) = (\pi_{ki}{}^A\beta\pi_{ik}{}^A)\pi_{kj}{}^B(X) = \alpha(Y),\end{aligned}$$

hence, B, $\alpha(X)$, $\alpha(Y)$ are collinear. ▎

The composite function $\pi_{ij}{}^B\pi_{jk}{}^A\pi_{kl}{}^B\pi_{li}{}^A$ is a permutation of $\tilde{m}_i - \{P\}$. If for every i, j, k, l the only permutation of this form that has a fixed point is the identity permutation, we say that P satisfies *condition* (G) for A, B. This is the algebraic form of a geometrical condition due to K. Reidemeister (Exercise 3).

LEMMA 3 **P** satisfies condition (G) for A, B if and only if $J_i{}^{AB}$ is a group for all i.

PROOF. Suppose **P** satisfies (G) for A, B. Let $\alpha = \pi_{ik}{}^A\pi_{ki}{}^B$ and $\beta = \pi_{il}{}^A\pi_{li}{}^B$ be elements of $J_i{}^{AB}$. Let $Q \in m_i$, $Q \neq P$. Then $R = \alpha\beta^{-1}(Q) \in m_i$. By Lemma 1 there is a

$$\gamma = \pi_{ij}{}^A\pi_{ji}{}^B \text{ in } J_i{}^{AB}$$

such that $\gamma(Q) = R$. Consider

$$\gamma^{-1}\alpha\beta^{-1} = \pi_{ij}{}^B\pi_{ji}{}^A\pi_{ik}{}^A\pi_{ki}{}^B\pi_{il}{}^B\pi_{li}{}^A = \pi_{ij}{}^B\pi_{jk}{}^A\pi_{kl}{}^B\pi_{li}{}^A.$$

Clearly, $\gamma^{-1}\alpha\beta^{-1}(Q) = Q$; hence, by (G) $\gamma^{-1}\alpha\beta^{-1} = 1$. Then $\alpha\beta^{-1} = \gamma \in J_i{}^{AB}$, and so (Section I.11) $J_i{}^{AB}$ is a group. Conversely, suppose $J_i{}^{AB}$ is a group. Suppose

$$\alpha = \pi_{ij}{}^B\pi_{jk}{}^A\pi_{kl}{}^B\pi_{li}{}^A$$

has a fixed point Q different from P. We have

$$\alpha = (\pi_{ij}{}^A\pi_{ji}{}^B)^{-1}(\pi_{ik}{}^A\pi_{ki}{}^B)(\pi_{il}{}^A\pi_{li}{}^B)^{-1} \in J_i{}^{AB},$$

since $J_i{}^{AB}$ is a group. α fixes Q, 1 fixes Q, so by Lemma 1, $\alpha = 1$. This proves (G). ▎

LEMMA 4 Let A, B, C be finite subgroups of a group G such that $|A| = |B|$. Suppose the elements of A and B can be listed $A = \{\alpha_i\}$, $B = \{\beta_i\}$ in such a way that $C = \{\alpha_i\beta_i\}$. Then $AB = BA$.

PROOF. Given any i and j, the product $\alpha_i\beta_i\alpha_j\beta_j$ is in C; hence, $\alpha_i\beta_i\alpha_j\beta_j = \alpha_k\beta_k$ for some k. Then $\beta_i\alpha_j = (\alpha_i^{-1}\alpha_k)(\beta_k\beta_j^{-1}) \in AB$, so $BA \subset AB$. Taking inverses, we also have $AB \subset BA$. ▎

LEMMA 5 Let $A_1, A_2, \cdots, A_n$ be finite subgroups of a group G such that $A_iA_j = A_jA_i$ for all i and j. Then $B = A_1A_2 \cdots A_n$ is a finite subgroup of G and every prime divisor of $|B|$ divides some $|A_i|$.

PROOF. We proceed by induction on n. The result is trivial for $n = 1$; suppose it holds for $n = k$. Let $A_1, A_2, \cdots, A_{k+1}$ be finite subgroups of G such that $A_iA_j = A_jA_i$ for all i and j. Let $B_1 = A_1A_2 \cdots A_k$, $B_2 = A_{k+1}$, and $B = B_1B_2$. Then $B_1B_2 = B_2B_1$. Also, by the inductive hypothesis, B_1 is a finite group satisfying the conclusions of the lemma. It suffices to show that B is a finite group and that every prime divisor of $|B|$ divides $|B_1|$ or $|B_2|$. Denote elements of B_i by letters with subscript i. Let $\alpha_1\alpha_2$, $\beta_1\beta_2$ be elements of B_1B_2. Then $(\alpha_1\alpha_2)(\beta_1\beta_2)^{-1} = \alpha_1((\alpha_2\beta_2^{-1})\beta_1^{-1}) = \alpha_1(\gamma_1\gamma_2)$, (since $B_1B_2 = B_2B_1$) $= (\alpha_1\gamma_1)\gamma_2 \in B_1B_2$. Hence, B is a group. By Section I.11, $|B|\cdot|B_1 \cap B_2| = |B_1|\cdot|B_2|$. Therefore, $|B|$ is finite and every prime divisor of $|B|$ divides $|B_1|$ or $|B_2|$. ▎

The next lemma has enough interest in itself to be called a theorem.

THEOREM 2 If **P** satisfies (G) for all A, B and all choices of P and l, and if n (the order of **P**) is a prime power, then **P** is Desarguesian.

PROOF. Let l be any line and P any point on l; let A, B, C be other points on l. Since J_i^{AB} is a group (Lemma 3), $J_i^{AB} = J_i^{BA}$. Let $\alpha_j = \pi_{ij}{}^B\pi_{ji}{}^A$ and $\beta_j = \pi_{ij}{}^A\pi_{ji}{}^C$ for $i, j = 1, 2, \cdots, n$. Then $\{\alpha_j\beta_j\} = J_i^{BC}$ and so, by Lemma 4, $J_i^{AB}J_i^{AC} = J_i^{AC}J_i^{AB}$. Let $H = J_i^{AB_1}J_i^{AB_2} \cdots J_i^{AB_{n-1}}$, where $B_1, B_2, \cdots, B_{n-1}$ are the points of l different from P and A. By Lemma 5 H is a finite group. Now $|J_i^{AB_k}| = n = p^r$ for all k, where p is a prime. Since every prime divisor of $|H|$ divides some $|J_i^{AB_k}|$, we must have $|H|$ a power of p, that is, H is a p-group. By Lemma I.11.2 then, H contains an element $\gamma \neq 1$ that commutes with every element in H. Since, as has been shown above, $J_i^{BC} \subset J_i^{AB}J_i^{BC}$, γ commutes with every element of every J_i^{BC}. Then, by Lemma 2, γ corresponds to a nonidentity element of $E(P, l)$. Then, by Theorem 3.3, **P** is Desarguesian. ▎

LEMMA 6 If **P** is Fano, then every nonidentity element of every J_i^{AB} is of order two.

PROOF. Let X be any point different from P on m_i. The diagonal points of the quadrangle P, B, $\pi_{ji}{}^B(X)$, $\pi_{ij}{}^A\pi_{ij}{}^B(X)$ are A, X, and $Y = \pi_{ji}{}^B\pi_{ij}{}^A\pi_{ji}{}^B(X)$. By the Fano property, A, X, and Y are collinear. Then $X = m_i \wedge AY = \pi_{ij}{}^A\pi_{ji}{}^B\pi_{ij}{}^A\pi_{ji}{}^B(X) = (\pi_{ij}{}^A\pi_{ji}{}^B)^2(X)$, and so $(\pi_{ij}{}^A\pi_{ji}{}^B)^2 = 1$. ▎

LEMMA 7 If **P** is Fano, then any two elements of any J_i^{AB} commute.

PROOF. Let $\alpha = \pi_{ij}{}^A\pi_{ji}{}^B$ and $\beta = \pi_{ik}{}^A\pi_{ki}{}^B$. By Lemma 6, $\alpha = \alpha^{-1}$

and $\beta = \beta^{-1}$. Then $(\alpha\beta)^2 = \alpha\beta^{-1}\alpha\beta^{-1} = \pi_{ij}{}^A\pi_{ji}{}^B\pi_{ik}{}^B\pi_{ki}{}^A\pi_{ij}{}^A\pi_{ji}{}^B\pi_{ik}{}^B\pi_{ki}{}^A = \pi_{ij}{}^A\pi_{jk}{}^B\pi_{kj}{}^A\pi_{jk}{}^B\pi_{ki}{}^A(\pi_{ij}{}^A\pi_{ji}{}^A)$ (we have inserted this last factor, which is just 1) $= \pi_{ij}{}^A(\pi_{jk}{}^B\pi_{kj}{}^A)^2\pi_{ji}{}^A = \pi_{ij}{}^A\pi_{ji}{}^A$ (by Lemma 6) $= 1$. Then, since $(\alpha\beta)^2 = 1$, we have $\alpha\beta = (\alpha\beta)^{-1} = \beta^{-1}\alpha^{-1} = \beta\alpha$. ∎

At last we reach the proof.

PROOF OF THEOREM 1. Each J_i^{AB} is a set of commuting permutations of $\tilde{m}_i - \{P\}$ (Lemma 7), which is transitive on $\tilde{m}_i - \{P\}$ (Lemma 1). Therefore, the group G_i^{AB} generated by J_i^{AB} is Abelian and transitive on $\tilde{m}_i - \{P\}$. Then, by Lemma I.12.2, G_i^{AB} is simply transitive on $\tilde{m}_i - \{P\}$. Hence, $G_i^{AB} = J_i^{AB}$, that is, J_i^{AB} is a group. Then **P** satisfies condition (*G*), by Lemma 3. The order of **P** is $|J_i^{AB}|$ which, by Lemma 6 and Lemma I.11.1, is a power of two. Then, by Theorem 2, **P** is Desarguesian. ∎

Combining this with Theorem IV.3.1, we have a corollary.

COROLLARY A finite plane is Fano if and only if it is isomorphic to some $\mathbf{P}_2F$, where F is a field of characteristic two.

Exercises

1. Prove Lemma 1.
2. Prove that the function $\eta \to \bar{\eta}$ of Lemma 2 is one–one.
3. Using Figure VI.5.1, find a geometrical interpretation of condition (*G*).

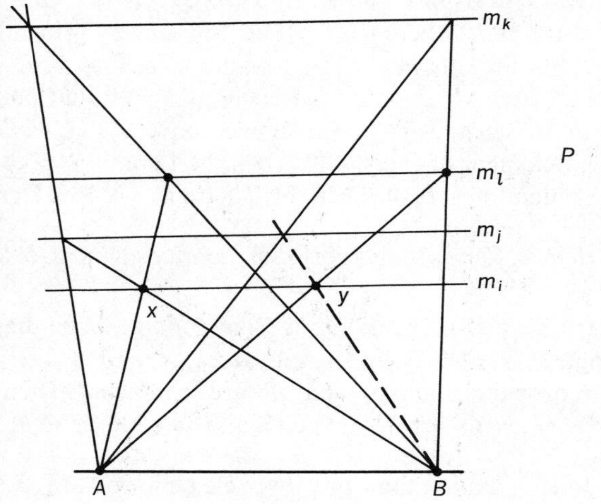

Fig. VI.5.1.

4. Assuming condition (G) for A, B, l, P whenever m_i, m_j, m_k, m_1 are distinct, show that the condition still holds when these lines are not distinct.
*5. Show that every Desarguesian plane satisfies condition (G).

6. CHANGE OF COORDINATES

The procedure explained in Section III.1 for coordinatizing an arbitrary plane depends, as do all such procedures, on an initial choice of certain lines or points. To what extent do the algebraic systems obtained in Chapter III depend on this choice? In this final section we shall discuss some results connected with this question.

We know from Section III.5 that if some planar ternary ring of a plane **P** is linear and is a field, then **P** is Pappian, and hence every ternary ring of **P** is linear and gives the same field. With a little care about right- and left-hand multiplication, we can replace "field" and "Pappian" in this statement by "division ring" and "Desarguesian," respectively. However, the word "linear" cannot be omitted: G. Pickert has given an example [17] of a non-Desarguesian plane some of whose planar ternary rings, although not linear, are fields.

Of course if the entire geometric structure of the plane can be carried over from one coordinate system to another, then the corresponding planar ternary rings are the same. This idea was essentially present in the proof of Theorem III.5.4. We make it precise in the following definition and theorem.

DEFINITION 1 Two planar ternary rings (R, T) and (R', T') associated with a given plane are *isomorphic* if there exists a one-one function α from R onto R' such that, for all x, y, z in R, $(\mathsf{T}(x, y, z))\alpha = \mathsf{T}'(x\alpha, y\alpha, z\alpha)$.

THEOREM 1 Let (R, T) and (R', T') be planar ternary rings associated with a given plane. (All elements of R' will carry primes.) (R, T) and (R', T') are isomorphic if and only if there exists a collineation μ of the plane such that $(0, 0)\mu = (0', 0')$, $(1, 0)\mu = (1', 0')$, $(1)\mu = (1')$, and $(\infty)\mu = (\infty')$.

PROOF. Suppose such a μ exists. Define α: $R \to R'$ by $(x, 0)\mu = (x\alpha, 0')$. (Is this all right?) Then $0\alpha = 0'$ and, for all x, y, z, $(x, y)\mu = (x\alpha, y\alpha)$ and $(z)\mu = (z\alpha)$. Then $(0', (\mathsf{T}(x, y, z))\alpha) = (0, \mathsf{T}(x, y, z))\mu = ((x, y)(z) \wedge [0])\mu = (x\alpha, y\alpha)(z\alpha) \wedge [0'] = (0', \mathsf{T}'(x\alpha, y\alpha, z\alpha))$, and so α is the required isomorphism.

Now suppose such an isomorphism α exists. Define μ, using the coordinate systems of (R, T) and (R', T'), as follows. $(x, y)\mu = (x\alpha, y\alpha)$, $(z)\mu = (z\alpha)$,

$(\infty)\mu = (\infty')$, $[m, x]\mu = [m\alpha, x\alpha]$, $[k]\mu = [k\alpha]$, $l_\infty\mu = l'_\infty$. Then μ is the required collineation. ∎

One of the best results on change of coordinates is that of L. A. Skornyakov [21]. He shows that if two different coordinatizations of a given plane each yield linear planar ternary rings that are V–W systems (Definition III.4.1), then these systems, although not necessarily isomorphic, have similar structures. Skornyakov's planar ternary ring is not ours; but, since ours is a special case of his (Exercise 6), this result holds for us.

We shall not give the rather lengthy definitions required to make precise the words "similar structures" in Skornyakov's result. Instead, we prove a simpler theorem of the same general type.

DEFINITION 2 Two binary systems $(R, \circ)$ and $(S, *)$ are *isotopic* if there exist three one–one functions f, g, h from R onto S such that for all x, y in R

$$f(x \circ y) = g(x) * h(y).$$

For properties of isotopy, see the exercises.

THEOREM 2 If two coordinatizations of a plane, of the type described in Section III.1, use the same points for X, Y, and I, then the additive loops obtained from the resulting planar ternary rings are isotopic.

PROOF. It is convenient to denote the unique solution in an additive loop $(R, +)$ of the equation $a = b + x$ by $x = (-b + a)$. $-b$ alone has no meaning.

Call the two coordinatizations the *old* and the *new*. We shall write new coordinates in ⟨sharp parentheses⟩. The same set R will be used for both planar ternary rings. We are given that $\langle 0\rangle = (0)$, $\langle 1\rangle = (1)$, and $\langle\infty\rangle = (\infty)$. Let the new origin O':$\langle 0, 0\rangle$ have old coordinates (a, b). Since the new x-axis passes through (0), each point $\langle x, 0\rangle$ has old coordinate of the form $(g(x), b)$. This defines a one–one function g from R onto itself. In Figure VI.6.1, let U:$\langle x, 0\rangle$:$(g(x), b)$, V:$\langle y, 0\rangle$:$(g(y), b)$. The figure shows the construction of Z:$\langle x \oplus y, 0\rangle$:$(g(x \oplus y), b)$, where $\oplus$ denotes addition in the new planar ternary ring. Let P:(a, w). Since V, P, I are collinear, $g(y) + b = a + v$, and so $v = (-a + (g(y) + b))$. Then, Q:$(g(x), (-a + (g(y) + b))$. Since Z, Q, I are collinear,

$$g(x \oplus y) + b = g(x) + (-a + (g(y) + b)). \tag{*}$$

Let $f(t) = g(t) + b$ and $h(t) = (-a + f(t))$. Since g is one–one onto and since $(R, +)$ and $(R, \oplus)$ are loops, f and h are one–one onto. With this notation (*) reads $f(x \oplus y) = g(x) + h(y)$; hence, $(R, +)$ and $(R, \oplus)$ are isotopic. ∎

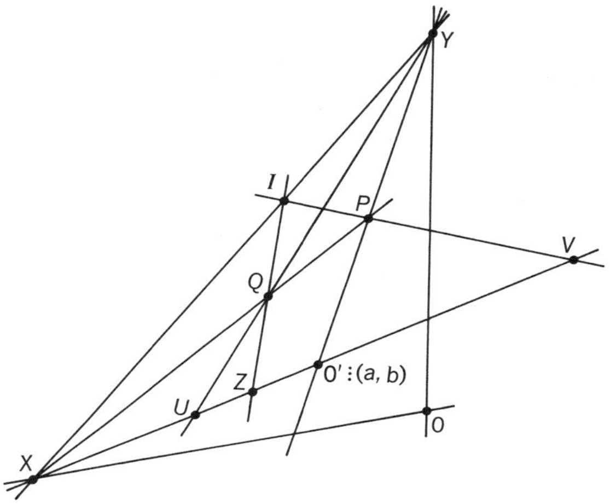

Fig. VI.6.1

Exercises

1. Under what conditions is an isotopy an isomorphism?
2. Prove that isotopy is an equivalence relation.
3. Show that in any field addition and subtraction are isotopic.
4. Are addition and multiplication isotopic as operations on the set of positive real numbers? As operations on the set of all real numbers?
5. Find and prove a result similar to Theorem 1 for the multiplicative loop of a planar ternary ring.
6. Skornyakov's introduction of a ternary operation in a plane **P** runs as follows: Let X, Y, O be noncollinear points of **P**. Let R be a set of cardinality the order of **P** that contains 0. Label $\tilde{X} - \{XY\}$ with R, being sure to label XO with 0; label $\tilde{Y} - \{YX\}$ with R, being sure to label YO with 0; label $\tilde{O} - \{OX\}$ with R, being sure to label OY with 0. Let a, b, c be in R. Let x_a be the line in $\tilde{X}$ with label a; z_b be the line in $\tilde{O}$ with label b; y_c be the line in $\tilde{Y}$ with label c. Define $\mathsf{T}'(a, b, c)$ as the label in $\tilde{Y}$ of the line $[(AB \wedge z_b)(AO \wedge y_c) \wedge x_a]B$.
   (a) Show how our ternary operation can be obtained by a special case of this method.
   (b) What general properties does the Skornyakov ternary operation have? (Compare Theorem III.1.2.)

Bibliography

1. Apostol, T. M., *Mathematical Analysis.* Reading: Addison Wesley, 1964.
2. Artin, E., *Algebraic Geometry.* New York: New York University lecture notes, 1958.
3. Baer, R., *Linear Algebra and Projective Geometry.* New York: Academic Press, 1952.
4. Barlotti, A., "Le possibili configurazioni del sistema delle coppie punto-retta (A, a) per cui un piano grafico resulta (A, a)-transitivo," *Boll. U.M.I.* (3), vol. 12, pp. 212–226 (1957).
5. Coxeter, H. S. M., *Projective Geometry.* New York: Blaisdell, 1964.
6. Coxeter, H. S. M., *The Real Projective Plane.* New York: McGraw-Hill, 1949.
7. Ver Eecke, P. (translator and editor), *Pappus d'Alexandrie. La collection mathematique* (two volumes). Paris, 1933.
8. Eves, H., *A Survey of Geometry* (two volumes). Boston: Allyn and Bacon, 1965.
9. Grothendiek, A., and Dieudonne, J., *Eléments de Géometrie Algébrique.* Paris: Publ. Math. de l'Institute des Hautes Etudes, still appearing in chapters.
10. Hall, M., *Projective Planes and Related Topics.* Pasadena: California Institute of Technology, 1954.
11. Hilbert, D., *The Foundations of Geometry* (translated by E. J. Townsend). Chicago: Open Court, 1902.
12. Jenner, W., *Rudiments of Algebraic Geometry.* New York: Oxford University Press, 1963.
13. Lenz, H., "Kleiner Desarguesscher Satz und Dualität in projektiven Ebenen," *Jahresbericht der Deutschen Math.-Ver.*, vol. 57, pp. 20–31 (1954).
14. Levy, H., *Projective and Related Geometries.* New York: Macmillan, 1964.
15. Lüneburg, H., "Endliche projektive Ebenen von Lenz-Barlotti Typ I–6," *Abh. Math. Sem. Univ. Hamburg*, vol. 27, pp. 75–79 (1964).

16. Ostrom, T. G., and Wagner, A., "On projective and affine planes with transitive collineation groups," *Math. Zeit.*, vol. 71, pp. 186–199 (1959).
17. Pickert, G., "Eine nichtdesarguesche Ebene mit einen Korper als Koordinatenbereich," *Publ. Math. Debrecen*, vol. 4, pp. 157–160 (1956).
18. Schellekens, G. J., *Geometries and Linear Groups*. New Haven: Yale University lecture notes, 1963.
19. Semple and Kneebone, *Algebraic Projective Geometry*. Oxford: Clarendon Press, 1952.
20. Semple, J. G., and Roth, L., *Introduction to Algebraic Geometry*. Oxford: Clarendon Press, 1949.
21. Skornyakov, L. A., "Natural domains of Veblen-Wedderburn projective planes," *Amer. Math. Soc. Translation*, series 1, no. 58 (1951).
22. Todd, R., *Projective and Analytical Geometry*. London: Pitman, 1947.
23. van der Waerden, B. L., *Modern Algebra*. New York: Ungar, 1953.
24. Walker, R. J., *Algebraic Curves*. New York: Dover, 1962.
25. Weil, A., *Foundations of Algebraic Geometry*. Amer. Math. Soc. Colloquium Publ., vol. 29, 1962.
26. Yaqub, J. C. D. S., "On projective planes of Lenz–Barlotti Class I6," *Math. Zeit.*, vol. 95, pp. 60–70 (1967).

BOOKS FOR FURTHER STUDY

Finally, we add a list of some of the most important recent books on projective geometry and related topics. Many of these have already been mentioned.

Albert, A. A., and Sandler, R., *An Introduction to Finite Projective Planes*. New York: Holt, Rinehart and Winston, 1968.

Artin, E., *Geometric Algebra*. New York: Interscience Publishers, 1957.

Artzy, R., *Linear Geometry*. Reading, Mass.: Addison-Wesley, 1965.

Baer, R.: [3].

Blumenthal, L. M., *A Modern View of Geometry*. San Francisco: Freeman, 1961.

Dembowski, H. P., *Finite Geometries*. Berlin: Springer, 1968.

Hall, M., *The Theory of Groups*. New York: Macmillan, 1959. (Chap. 20: Group Theory and Projective Planes)

Lenz, H., *Vorlesungen über projektive Geometrie*. Leipzig: Academische Verlagsgessellschaft Geest und Partig, 1965.

Levy, H.: [14].

Pickert, G., *Projektive Ebenen*. Berlin: Springer, 1955.

Walker, R. J.: [24].

Weir, A. J., and Gruenberg, K., *Linear Geometry*. Princeton, N. J.: Van Nostrand, 1967.

Notation

For reference style and logic, see Section 1 of Chapter I. For standard set theory notation, see Section 2 of Chapter I.

$\mathbf{A}_n F$, affine n-space over field F, 39
$A \subsetneq B$, A is a proper subset of B, 1
C, complex numbers, 4
$(\mathbf{P}^1\mathbf{P}^2\mathbf{P}^3\mathbf{P}^4)$, cross ratio of hyperplanes, 83
$(P_1P_2P_3P_4)$, cross ratio of points, 79
$(x_1x_2 \cdots x_n)^k$, cyclic permutation, 13
det A, determinant of matrix A, 6
I, end of proof, 1
$\mathbf{E}_n$, Euclidean n-space, 16
$p_{x^iy^j}$, formal partial derivative, 8
$(Q_0Q_1 \cdots Q_r|U)$, frame of reference, 78
$A \vdots \alpha$, "geometric object A has the coordinate α," 21
$[u_0, u_1, \cdots, u_n]$, hyperplane coordinate, 75
ϵ, incidence relation, 24
$a \equiv b (\mathrm{mod}\ m)$, integer a is congruent to b modulo m, 11
$m|n$ $(m\nmid n)$, integer m divides (does not divide) n, 11
$\mathbf{SS}'$, join of subspaces, 30
$[p, q, r]$, line coordinate, 21
$\mathbf{S} \wedge \mathbf{S}'$, meet of subspaces, 30
$\mathbf{P}_n(P, \mathbf{P}_{n-1})$, n-dimensional projective subspace, 28
$|A|$, number of elements in finite set A, 2
$\tilde{l}$, range with axis l, 24
$\tilde{P}$, pencil with center P, 24
$H(AB, CD)$, points A and B separate harmonically C and D, 82
$\mathbf{P}_n F$, projective n-space over a field F, 30
$\langle P_1, P_2, \cdots P_r\rangle$, projective order, 94
R, real numbers, 4
$R(p, q)$, resultant of polynomials p and q, 9
$[M]$, subspace spanned by M, 30
T, ternary operation of planar ternary ring, 45
A^T, transpose of matrix A, 6

Index